FOUNDRY ENGINEERING

FOUNDRY ENGINEERING

HOWARD F. TAYLOR · MERTON C. FLEMINGS

JOHN WULFF

DEPARTMENT OF METALLURGY
MASSACHUSETTS INSTITUTE OF TECHNOLOGY

Illustrations by George E. Schmidt, Jr.

NEW YORK · JOHN WILEY & SONS, INC.
LONDON · CHAPMAN & HALL, LIMITED

Library of Congress Catalog Card Number: 59–11811

Preface

Since the first foundry center came into being, in the days of the Shang dynasty (1766–1122 B.C.) in China, metal casting has traditionally been an art and a craft with secrets of the trade passed jealously from father to son. Only in the last century have science and engineering made noticeable inroads on the materials and processes of the foundryman. But casting will always be one of the most economical routes from raw material to finished metal products, and it was inevitable the art of the founder would yield to the economy and precision of the engineering approach.

This book is concerned with the basic engineering aspects of the materials and methods involved in making castings. It is not a reference book in that it does not deal with minute details of processes nor give elaborate recipes or properties of materials. It is a simplified, but comprehensive, engineering description of the foundry and of founding presented in a straightforward teachable fashion.

After several years' experience in teaching graduates and undergraduates at M.I.T., the authors have learned it is not easy to give a student of engineering a "feeling" for metal casting or an appreciation of the problems of the foundry industry without an early accurate understanding of the manufacturing processes involved; this understanding, once attained, seems also to enhance interest and enthusiasm and makes the student eager to tackle the more complex scientific problems in theses and special researches. To achieve the above ends is the purpose of this book; only the most easily understood descriptions and simplified sketches are used to explain what a casting is, how casting is done, and what unique engineering and metallurgical properties and economies make castings an important material of construction.

Although this book is intended primarily as an introductory college text, an able teacher should find it suitable for use in high schools and vocational schools, and for in-plant training and reference. A one-term college course of two one-hour lectures and four hours' homework per week is sufficient, we have found, to cover the material presented in this book whether the student is a freshman or a senior. In high school or vocational school courses somewhat more time may be spent usefully in acquiring the neces-

sary basic science background for more complete understanding
of Chapters 4, 5, 8, and 14, although none of the concepts are
more difficult than high school mathematics, chemistry, or physics.

An earnest attempt has been made in the writing of this book
to preserve as much of the arts and crafts flavor of the skilled
foundryman as is possible in a textbook devoted to engineering
principles.   Even so, no matter how well this is done, the student
will find it necessary to spend time in a real foundry or a foundry
laboratory to fully appreciate the fascination of handling and pouring molten metal into beautiful art objects or into engineering parts
which will perform dependably in service.

HOWARD F. TAYLOR
MERTON C. FLEMINGS
JOHN WULFF

August 1959

Contents

CHAPTER 1

Introduction to Casting Processes

1.1 Introduction to casting and to castings

In the broad language of engineers, useful engineering parts are made from metals and alloys by *metal forming*. Many ways of forming metals are available; they include *casting, forging, welding, electroforming, powder metallurgy,* or a combination of these methods (e.g., as in a cast-weldment in which parts of the final item are cast and welded together or are welded to other parts made by forging). In addition, whatever processes are used for a given part, machining (metal cutting) may be required to a greater or lesser degree.

This book treats only the *casting (founding, foundry)* aspects of metals fabrication, and attempts to delineate the process as a basic branch of industry, as well as to describe what a casting is, how it is made by different specific processes, and what unique characteristics castings possess as engineering material. It is hoped that this knowledge will bring about a more intelligent and broader use of castings, improved designs, and more dependable engineering production.

Casting, as a method for shaping metals, antedates 4000 B.C., when arrowheads of copper were found to be more easily poured in molds (cast) than they could be beaten (forged) to size and shape. The genesis of early foundry development has been well treated by Simpson [1] in his *Development of the Metal Casting Industry*.

From such an early beginning, it might be expected that the foundry industry would be among the first branches of engineering to shed its cloak as an art and emerge as a full-blown science; unfortunately, this did not occur. Before the advent of machine methods, a high order of skill was required to make good *molds* for shaping the molten metal required for casting; accordingly, craft guilds, or closely guarded family groups (perpetuating father-to-son secrets), kept metal casting a closely guarded art.   Indeed, it was only after the turn of the twentieth century that intershop visitations were encouraged or permitted; even today some shop practices are closely guarded secrets.   In 1540 Biringuccio was much in disfavor with his fellow craftsmen for publishing a book [2] describing the so-called God-given art of the foundryman.   Today statuary casting is fading rapidly as an art in America because the sons of the craftsmen are not interested in following in their fathers' footsteps, and the fathers are not willing to share their secrets freely with others.

The middle of the twentieth century saw a marked renaissance in the casting industry.   A number of fortunate developments combined to bring it about: (1) The average age of personnel in the administrative bracket of the foundry industry exceeded 60 years, and young men of ability were not being trained; (2) rapid development in other branches of the metal processing industry seriously threatened the future of foundries unless bright young engineers could be interested in the field as a profession; (3) the Foundry Educational Foundation was formed by a coalition of technical and trade associations representing the foundry industry with the aim of aiding and encouraging foundry training at the college and university level; and (4) colleges and universities began to teach foundry engineering on an enlightened and advanced scale.

So the second half of the twentieth century is seeing casting processes gradually lifted from the art or crafts status to that of a science —engineers are complementing and supplementing the craftsman, and for the first time books are being published treating casting processes on the level of an advanced engineering or scientific subject; for, contrary to the opinions of many, no other branch of industrial engineering will lend itself more responsively or rewardingly to scientific treatment and control, although the road will not be easy because the start has been slow and the numbers of scientists and engineers engaged in the process are relatively few.

This book will not attempt to delineate or develop the basic sciences underlying foundry processes; instead it will deal with the how and why of these processes as a foundation stone for more advanced books; it is, in short, *an elementary, more or less descriptive, engineering textbook for use at the college or university level.*

1.2 Further introduction to founding for the complete novice

After several years of attempting to teach foundry practice, it has been impressed on the authors' minds that one of the most important problems in this field is to reach an early, oversimplified understanding of what a casting is and how it is made. To accomplish this, we will use (1) an arrowhead, which represents the simplest of castings, (2) an integrally cast cup, spoon, and saucer which has been used for many years as the supreme test of a molder's skill, and (3) a single segment of a household radiator. Among these extremes lies the whole gamut of foundry practice, whether the job is to be done by hand or machine. If we clearly understand each step in making these three items, we can proceed with clarity to the myriad engineering details involved in foundry practices.

One needs only a few simple things, and limited skill, to make an arrowhead—a pattern, material for a mold, a pot in which to melt some copper, and a fire (for melting the copper). Going back 6000 years, one would find (1) a stone arrowhead to use as a pattern, (2) clay or stone, or indeed a slab of precast copper, to use as a mold, (3) a clay or stone pot for melting, and (4) a wood fire, fanned or blown by crude bellows to make it give enough heat for melting copper. If stone is used as molding material, a pattern is not needed—a cavity the shape of the arrowhead desired is chipped into the stone with flint or some other hard, sharp object. If clay is used a soft, plastic mass is first formed by wetting the clay with water from a nearby brook and kneading it with the fingers. A small mass is laid upon a piece of stone, or the ground, and the top flattened. The arrowhead pattern, preformed from stone or wood, is then pressed into the flat top of this piece of clay only so far as to permit it to be withdrawn later without damaging the mold. A layer of fine, dry sand or other powder is sprinkled over the mold to prevent sticking; then another flat piece of clay, just like the first, is laid and pressed firmly over the first half of the mold which contains the pattern. The mold is baked slowly or dried in the sun; the halves are then separated, and the arrowhead pattern is removed.

All that now remains is to scratch a small channel through which the molten copper can flow into the mold cavity. The halves of the mold are clamped together and the mold is ready for use. The pot of clay or stone is filled with chunks of copper (or copper-bearing stone) and placed over a hot fire. As soon as the metal is melted and heated until it is fluid enough to flow through the small channel into the mold, the pot is removed from the fire and the metal poured—probably several molds would be used to take up the contents of a large pot. When the metal has cooled (frozen), the mold is pried apart and the cast arrowhead removed, ready for use after the pouring channel is broken from the casting. The stone mold is a permanent type and can be reused, but the clay mold is usually expendable, and a new one must be prepared for each casting. The steps in preparation of a simple arrowhead casting are sketched in Fig. 1.1.

Plastic clay is a satisfactory molding medium for very simple shapes such as the arrowhead; however, because it is difficult to work with and tends to crack on drying, the early artisan began looking for a better material. He found it in sand containing a small amount of clay as a natural binder. This molding sand, often simply called "sand," has been in use for thousands of years up to the present day. The complexity of our next example of the art of metal casting will require its use.

Figure 1.2 shows the other extreme of the molder's skill, and typifies the steps often required in making more intricate castings, whether by hand or by machine. Assume that we have been assigned the task of casting a saucer, cup, and spoon in one piece (if each piece were to be reproduced separately, the task would not be noticeably more difficult than making the mold for arrowheads). Let us assume also that we are to use clay-bonded sand as our molding medium. How to proceed? Would we want a one-piece or a three-piece pattern? How would we prepare our pattern for use? After some thought, it would be apparent that we would use a cup, spoon, and saucer from the kitchen, and we would *not* fasten them together. We would first lay the saucer upon a flat board and press sand firmly around it, as shown in Fig. 1.2b. Also shown in Fig. 1.2b is a conical-shaped piece of wood which is pressed into the sand over the saucer; the hole formed by this piece of wood will later serve as a flow channel for metal to enter the casting cavity. The next step in the molding process would be to turn the first sand segment over, set the cup on the saucer, and pack sand firmly around the cup; we must be sure to separate the two sand segments with a layer of dry powder, so that they can later be pried apart.

1 THE RAW MATERIAL FOR CASTING AN ARROWHEAD — CLAY, WATER, COPPER, AND STONE PATTERN.

2 THE CLAY IS TEMPERED WITH WATER AND KNEADED UNTIL PLASTIC.

3 THE PATTERN IS PRESSED INTO THE SURFACE OF THE CLAY.

4 THE SURFACE IS LIGHTLY DUSTED WITH FINE DRY SAND TO PREVENT STICKING.

5 ANOTHER LUMP OF CLAY IS PRESSED OVER THE FIRST TO ENCASE THE PATTERN.

6 AFTER THE MOLD HAS DRIED IN THE SUN, IT IS PRIED OPEN, THE PATTERN REMOVED, AND A POURING CHANNEL IS SCRATCHED IN THE CLAY.

7 FINALLY, MELTED COPPER IS POURED INTO THE MOLD. WHEN IT HAS COOLED, THE MOLD IS BROKEN OPEN – AND THE CASTING REMOVED.

FIG. 1.1. Steps in casting an arrowhead.

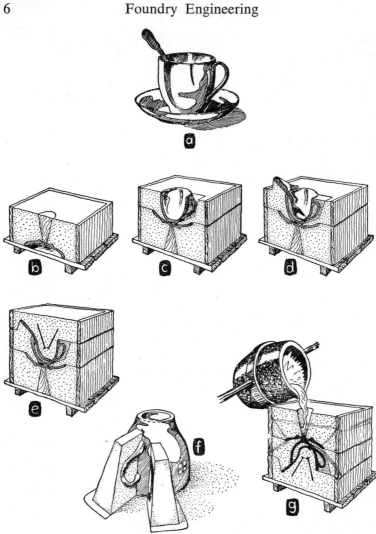

Fig. 1.2.   Molding a cup, spoon, saucer.

The molding problem now becomes noticeably more difficult—
the bowl-shaped sides of the cup, and the handle, make removal
of the pattern a tricky consideration.   Before proceeding further,
we must cut a hole in the sand around the handle (Fig. 1.2c) to
permit removal of the pattern in the vertical direction at the com-

pletion of molding. To make the impression of the handle, two
removable sand segments are formed in the cavity we have just
cut. First a dry powder is dusted in the cavity; then one segment
is formed by pressing sand into one side of the cavity, being careful
to end the segment in a sharp line *exactly* down the midsection of
the handle. Figure 1.2*d* illustrates the segments in place in the
mold, and Fig. 1.2*f* shows the segments as they look when withdrawn.

Next the cup is filled with sand to about where the spoon will
be in the cup, and the spoon pressed firmly into the sand up to the
exact center line of the handle and rim edge of the spoon (Fig.
1.2*d*). Dry "parting" powder is sprinkled over the top of the
mold and another mass of sand pressed over the top of the cup and
spoon (Fig. 1.2*e*). With this operation completed, the mold can
be dried or baked with the pattern in place. The mold parts are
then separated along the "parting lines" produced by the dry
powder, and the pattern is removed. Figure 1.2*f* shows the last
step in the operation of removing the pattern—the separation of
the small segments from around the handle. When all the mold
parts are reassembled, metal is poured through the flow channel
to fill the cavity and produce the casting (Fig. 1.2*g*).

By using the simple example, and oversimplified description, of
the molding and casting of an arrowhead, we are introduced to a
casting and to the art of the foundryman; by using the more complex example of the cup, spoon, and saucer, we see that molding
often requires careful planning and no little skill. We have learned
that (1) sand, clay, stone, or metal can be used as a molding
medium, (2) sand and clay molds are expendable, and stone or
metal molds can be used repeatedly, (3) some castings are simple
to make, but others are intricate, (4) making segments of a mold
and pouring the metal into the mold comprise the process of casting or founding, and the object so made is a casting.

Our third example of a casting—a segment of the old-fashioned
radiator for home heating—illustrates another point. Here we have
an example of a casting nearly all hole—so much hole, in fact,
that a complete novice might have trouble planning how to make
it. Molding here is essentially the same as for the arrowhead, so
that this presents no problem. To form the hole, a hard, baked
sand segment is used, held in place during casting by little metal
spacers. After the casting is poured, the sand segment breaks

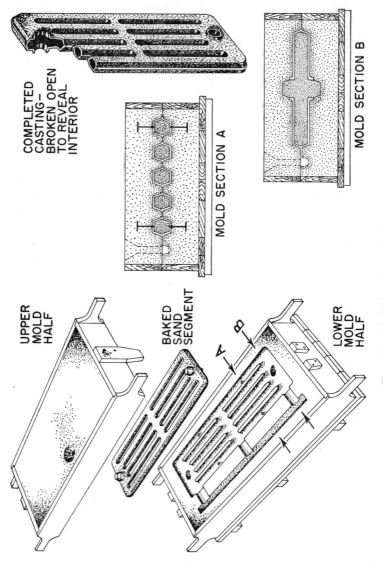

COMPLETED CASTING— BROKEN OPEN TO REVEAL INTERIOR

MOLD SECTION A

MOLD SECTION B

UPPER MOLD HALF

BAKED SAND SEGMENT

LOWER MOLD HALF

Fig. 1.3. Molding a household radiator.

down or crumbles because of the heat of the molten metal, and can be jiggled from the small holes at the ends of the casting. The process for molding a radiator is sketched in Fig. 1.3. The pattern which formed the external shape of the casting was a duplicate, or positive, reproduction of the piece to be made, but the internal segment (core) was made from a negative (corebox) of the form reproduced in metal. A little thought will be required to make this difference apparent.

1.3 A simplified description of foundry operations

Brief reference is made to Figs. 1.4 through 1.9 in order to orient certain aspects of the foundry industry before tracing its product from raw material to finished casting.

Figure 1.4 traces the growth of the industry since 1939, showing more than doubled capacity in less than 20 years. World War II is, of course, responsible for much of this expansion. It is significant, from Fig. 1.5 that gray iron accounts for most of the total casting tonnage (77 per cent). Figure 1.6 is added to put

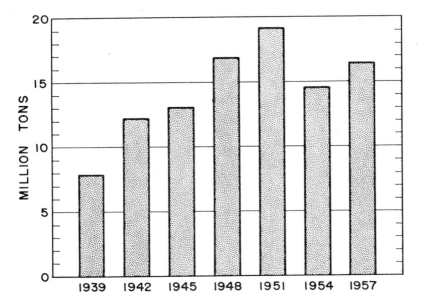

FIG. 1.4. Growth of the foundry industry, 1939–1957, showing tons of castings produced annually.

the light metals (aluminum and magnesium) in more comparable relationship to ferrous metals and bronzes; their relatively low density, compared to the ferrous metals, makes any production comparison on a basis of weight alone misleading. Also, light metals are relative newcomers to the foundry field. Copper-base alloys and gray iron have been cast for 6000 and 2500 years, respectively, whereas aluminum and magnesium castings have become commercial products only in the last half century. Another light metal, titanium, is still so young that only a very few castings (other than ingots) have yet been made. It is safe to predict that light metal castings will become steadily more important, and their volume in castings will increase several hundred per cent in the next ten years.

Figure 1.7 is included to show how the 5758 foundries in the United States are distributed as to size, as reflected by the numbers of employees. This chart indicates that the foundry industry is largely comprised of small businesses; only forty-three (43) foundries employ over 1000 men each, while 2937 are units of 20

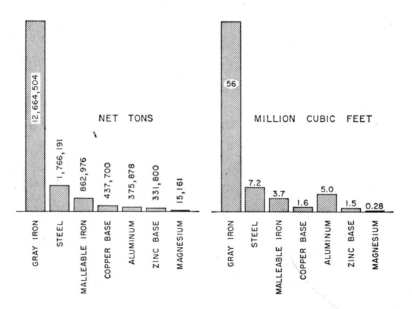

Fig. 1.5. Weight of castings produced, 1957 (in the United States).   Fig. 1.6. Volume of castings produced, 1957 (in the United States).

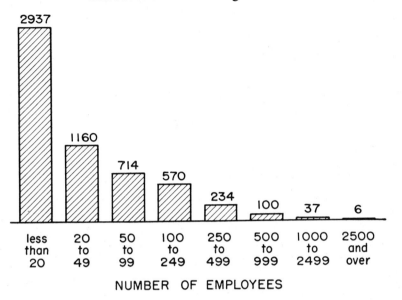

FIG. 1.7. Distribution of foundries in the United States according to number of employees, 1957. Vertical height represents number of foundries.

men or less. Figure 1.8 shows how the 5758 units are divided as to type of metal cast. Since some shops cast more than one metal, Figs. 1.7 and 1.8 will reflect some disparity. Figure 1.9 illustrates that the majority of the foundries are located in the east and middle west; only 25 per cent of the foundries are west of the Mississippi.

Figure 1.10 is a flow sheet of the manufacturing steps that comprise a typical foundry operation. A brief introduction to the individual steps required to make a casting from raw material to finished product will clearly delineate this field of metal fabrication from all others. Each step in the process is treated in detail in later chapters.

From the first half of this chapter we recall that sand, a pattern, and molten metal are basic requirements for a casting operation. In a well-run shop, the sand is tested for quality in a sand laboratory, and chemical analyses may, or may not, be made of the metal to be charged into the furnace. Usually, a good scrap yard foreman knows the composition of his scrap within close limits; for

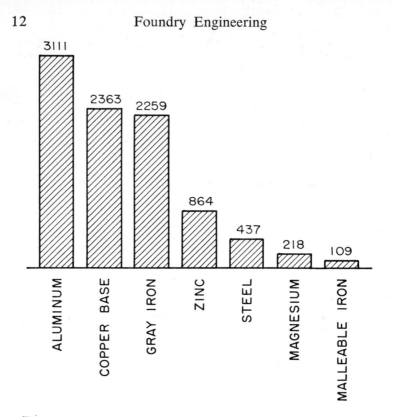

Fig. 1.8.   Distribution of foundries in the United States according to metal cast, 1957. Vertical height represents number of foundries.

example, stove plate scrap is made of high phosphorus iron, and worn or obsolete railroad side frames, sold as scrap, are of a closely specified grade of steel, etc. Pigs of secondary (remelted) metal may be used with or instead of scrap, and sold to definite specifications. If necessary, however, chemical analyses can be made of the metal to be used for melting.

Sand is blended with proper ingredients (water or oil and clay) in a suitable mixing device; the "mulled" sand may then be used for molding around the pattern, in which case clay is used as a binder or may be mixed with linseed oil for cores and later baked to develop strength. Scrap and/or pigged metal is fed into a furnace where it is melted; it is then poured from the furnace into ladles of convenient size and so transferred for pouring into molds; the molds may be set upon the floor, on roller conveyors, or brought to the ladle on turntables or cars.

After the molten metal is poured into molds and allowed to solidify, the sand is removed from the casting, usually in a mechanically vibrated "shakeout device," and cleaning operations are begun.   Gates (channels in the mold through which the metal flows to fill the cavity left by the pattern) and risers (metal reservoirs which feed molten metal into the casting as it freezes and shrinks) are removed by saw, torch, or other devices.   Chipping or grinding (snagging) away excess metal left after gates and risers are removed, is followed by sand blasting or by some other suitable cleaning process.   Welding may be required to repair defects, and this is often followed by heat treatment to remove stresses or improve properties.   The casting is inspected according to specifications for the particular metal used or part made, and is then ready for shipment.   More than one operation of cleaning, heat treatment, or inspection may be necessary, and various sequences may be used, before a casting is ready for shipment; also, specific operations vary widely for different metals but, in general, the operations will be as outlined above, and as shown in Fig. 1.10.

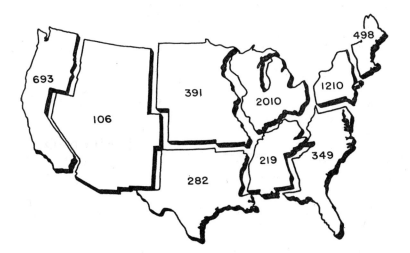

FIG. 1.9.   Geographical distribution of foundries in the United States, 1957.

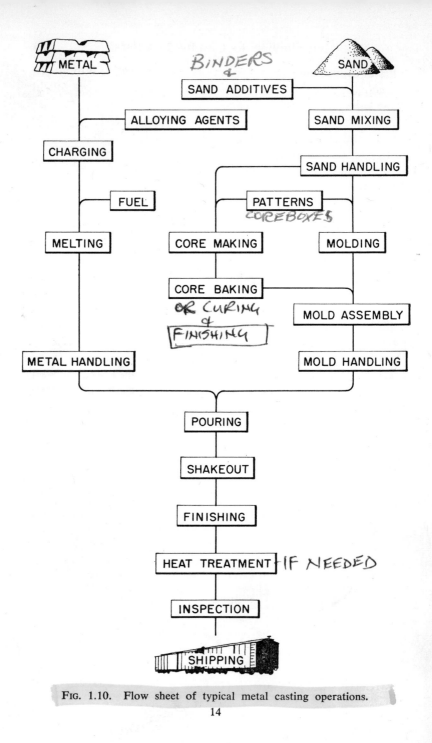

FIG. 1.10. Flow sheet of typical metal casting operations.

14

1.4 Summary

We now see that (1) simple, solid castings can be made from a simple pattern by pouring molten metal into a cavity left in a mold after removing a pattern, but (2) some castings require cores.

We have also had a brief introduction to the operations that comprise founding as a basic industry, and have taken a quick look at some foundry statistics. We are now ready to undertake a detailed study of specific operations.

Whether the casting is simple or complex, small or large, hand- or machine-molded, custom-made or made on the production line, or whether the casting is of silver, gold, bronze, or gray iron, the steps in its manufacture are essentially those described in the examples above. It is somewhat surprising that a technical, engineering treatment of these steps should require the many pages of text that follow; but so far we have considered only castings which do not require great strength, ductility, soundness, fine surface finish, or accurate dimension to perform their assignment satisfactorily. To incorporate these properties dependably and reproducibly in castings requires engineering skill and knowledge and careful control of operations—it is for the clarification of these ends this book is written.

1.5 Problems

1. What would be the advantages of a metal mold compared to one made with sand? What would be the disadvantages?

2. Why are most foundries in the United States small (shops less than 20 men) whereas most forging or metal rolling shops are large?

3. What is the difference between a pattern and a corebox? Between a pattern and a core? Are entire molds made of cores?

4. Why do you think sand is so widely used as a molding material? What other materials can you think of that might be used in its place?

5. List 5 problems you would anticipate in the melting of metals for casting?

6. Prepare a list of 25 well-known castings used for household, industrial, or art applications.

7. Some typical cast metals are silver, gold, tin, lead, platinum, pure iron, gray cast iron, pure copper, copper-tin alloys, pure aluminum, aluminum-silicon alloys, magnesium and plain carbon steel; arrange these in the order of their melting points (using any applicable reference such as this book, ASM *Handbook,* handbooks of chemistry and

physics, etc.) and list 2 typical castings that you think are made from each.

8. In Problem 7 see if you can predict which metals and alloys might be melted in (1) a Pyrex beaker, (2) a gas-fired crucible furnace, (3) an electric arc furnace; then check your answer in Chapter 12.

9. If you could make a casting of ice in a typical mold and one of pure iron (metal) to close dimensional accuracy, what differences would there have to be in the patterns used?

10. Calculate the upward force on the cope used in making a casting of pure iron 12 $\times$ 12 in. square and 1 in. thick, using a downgate 6 in. high entering at the bottom edge of the casting; ditto for aluminum.

1.6   Reference reading

1. Simpson, B., *Development of the Metal Castings Industry*, A.F.A., Chicago, 1948.

2. Biringuccio, Vannoccio, *De La Pirotechnia,* written in 1540 and translated in 1942 from the Italian by Cyril Stanley Smith and Martha Teach Gnudi; A.I.M.E., New York, 1942.

3. *Transactions,* A.F.S., Des Plaines, Ill.

4. *Proceedings, Institute of British Foundrymen,* Manchester, England.

5. *Modern Castings,* A.F.S., Des Plaines, Ill.

6. *Foundry,* Penton Publishing Co., Cleveland.

7. *Foundry Industry Marketing Guide,* Penton Publishing Co., Cleveland, 1957.

CHAPTER 2

Sand Casting

2.1 General engineering background

Although an ingot is a casting in every sense of the word, castings, as products of the foundry, are generally considered objects made as nearly to the shape in which they are to be used as possible. It is seldom that any additional forming process, such as forging, stamping, pressing, or coining, is performed on castings (other than ingots). Machining or grinding is usually necessary, but special casting processes can be controlled closely enough to yield castings not requiring finishing operations; investment casting is an example.

Castings are used because they have specific important engineering properties; these may be economic or metallurgical. Objects amenable to casting are, in general, cheaper than forgings or weldments because patterns for molding are usually less expensive than dies, jigs, and fixtures. Obviously this economy depends upon the quantity of parts to be made. If production is high enough to amortize equipment, and the part is equally amenable to manufacture by other methods, forging, welding, or some other means of fabrication may be indicated.

Properly made castings have characteristics of special interest to engineers. Castings do not have directional properties; no laminated or segregated structures exist, since metal is not displaced after solidification. This means that strength, ductility, and toughness are equal in all directions—an important consideration in such applications as gun tubes, cylinder liners, gears, piston rings, etc. Cast iron has damping characteristics unique among structural alloys; it is especially good for lathe beds, engine frames, and other applications where damaging vibrations prevail. The ability to flow molten metal into thin and intricate sections pro-

17

vides an important latitude in design of castings. Castings can be made of any metal or alloy that can be melted; this makes possible manufacture of objects of materials which cannot be otherwise shaped; jet turbine blades are an example. For many applications like tank armor, the streamlined casting is less vulnerable to shell penetration than fabricated units; also, streamlined designs are functional and esthetic.

2.2   Sand casting

Silica sand (SiO_2) is used more universally for making castings than any other molding material. It is relatively cheap, its base cost ranging from $2.00 to $10.00 per ton, and is sufficiently refractory even for steel foundry use. A suitable bonding agent (usually clay) is mixed or occurs naturally with the sand; the mixture is moistened with water to develop strength and plasticity of the clay and to make the aggregate suitable for molding. The resulting sand mixture is easily prepared and molded around various shapes to give satisfactory castings of almost any metal.

The fundamentals of mold making are simple, but expert hand molding requires much skill and practice. Production line work is done today by *machine molding,* in which nearly all operations are automatic. The skilled molder is replaced by a relatively untrained machine operator. A *mold-a-minute* is not an uncommon production rate.

The process of molding is shown schematically in Fig. 2.1. The simplest form, *hand molding,* will be described, since machine processes are essentially improvements upon this basic method. A few simple terms must first be established. *Sand* refers to the molding material, the simplest type being a mixture of silica sand, clay, and water, blended by hand or mechanical mixer. The *flask* is a frame of steel, iron, or wood in which the mold is made. The *cope* is the upper and the *drag* the lower half of the mold; intermediate sections, if used, are *cheeks.* The *rammer* is provided with a *peen* and *butt* end for ramming sand around the edges of the flask and over the pattern respectively. The *pattern* is the form around which sand is rammed; it may be of one piece or of split construction, depending upon size and configuration. *Bottom boards* are flat plates of wood or metal upon which patterns with at least one flat side or one-half of a split pattern is laid before *ramming* the drag.

MOLDING/ FOLLOW/ OR RAM UP BOARDS

As to the mold proper, the *mold cavity* is the empty shape left in sand when the pattern is removed and into which metal is poured, or *cast,* to form the *casting.* The mold cavity may not be left completely empty; a *core* of firmly baked sand may be used to form the internal shape and dimensions of the casting. *Chaplets* are metal objects for holding the core in position against the washing and lifting effect of molten metal. *Core prints* are other means for securing the core; these are extensions of the pattern which form a cavity in the mold into which identical extensions of the core are fitted. The *prints* are usually square or rectangular and are reasonably large as compared to the core to prevent turning or shifting. *Clamps,* or *weights,* are used for holding the parts of the flask together.

Molten metal is introduced into the mold cavity through the *gating system,* which includes *downgate* (*sprue*), *crossgate* (if used), and *ingate.* A *pouring cup* or *pouring basin* is either placed upon or formed into the upper cope surface, connecting to the downgate, to receive the metal from the ladle.

In making a simple mold (Fig. 2.1), the bottom board is first laid securely on the bench or floor. The pattern is positioned, the flask is located properly, and the sand is sieved over the pattern with a *riddle.* The sand employed for this purpose will touch the metal; in the best practice, it is an especially fine and clean variety of sand, called *facing sand.* Used *backing sand* may then be shoveled into the flask over the facing, and the whole rammed securely, using the peen around the edges and the butt rammer over the pattern. Once the flask is rammed properly a straightedge, or *strike,* is drawn across the upper surface to remove excess sand; a second flat board is placed upside down on the mold and flask and clamped or held in position while the whole is overturned. The original bottom board, which is now on top, is removed, and the drag half of the mold is essentially finished. It will later be necessary to cut an ingate to the mold cavity unless one is provided as part of the pattern, *as is the recommended practice.* It may be necessary to *tool* the drag somewhat, i.e., smooth the sand around the pattern edges or otherwise shape it. *Parting powder,* a white powdery substance, is dusted over the sand and pattern, and the cope flask is positioned. If patterns are provided with downgates and crossgates, these are set in place, as are the patterns for whatever risers are used. Downgates are sometimes

[handwritten margin notes: "OR RUNNER", "RAM-UP BOARD", "FOLLOW BOARD OR MOLDING BOARD", "LAST RAM ONLY", "NOW BOTTOM BOARD"]

cut through the cope with a cylindrical tube after the cope is finished. A layer of facing sand is riddled over the drag and pattern as before, and backing sand is added and rammed solidly. A steel bar is again used to *strike off* the upper surface of the

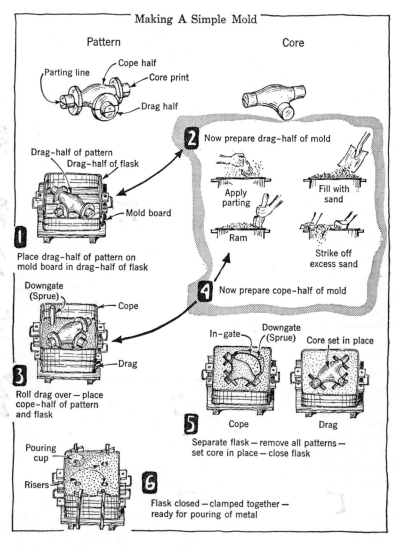

Fig. 2.1.

cope; sprue and riser patterns are then *drawn*. The cope is removed from the drag and usually placed aside for any finishing
operations required. If split patterns are used one-half will be
picked up with the cope; so the cope is laid upside down on a
spare bottom board. The patterns are then *rapped* slightly to
loosen them from the sand and are removed from cope and drag
with a *draw spike;* ingates are cut, the mold is patched as required,
chaplets and cores are positioned, and the mold is *closed* by placing
the cope again atop the drag. Clamps or weights are placed on the
flask to resist the tendency of the cope to *float* or *shift* as molten
metal fills the mold cavity.

Sometimes a *snap flask* is used, which is a single flask used
repeatedly. As soon as the mold is closed and positioned for
pouring, snaps at one edge of the flask are loosened, the sides are
pulled back on hinges, and the flask frame is entirely removed. A
jacket and weights are placed on the mold before pouring, and the
flask is ready to use again.

2.3   Sand-clay molding mixtures

Sands and clays are formed by chemical and mechanical weathering of rocks; deposits are found wherever left by wind, water, or
glacial action. Frequently sands are found in intimate contact
with clay, and in these cases the sand-clay mixtures may be used
essentially as mined—the molder has only to mix water with the
aggregate and the clay develops the strength and plasticity for
molding. Such sands are termed *natural sands*.

Sands are also found washed free of clay by nature; or the sand
may be purposely separated from the clay in large tanks by pro-
ducers of foundry sand. These sands must be blended, before use,
with clay and such other materials as desired or required; they
are designated *synthetic sands*. The term "synthetic" is somewhat
of a misnomer since it is not the sand, per se, that is synthesized,
but the sand-clay mixture however, this term has been firmly established in shop vernacular by common use.

Natural and synthetic sands have unique differences. *Natural
sands* are used as mined, may contain considerable organic matter
and from 5 to 20 per cent clay, and require 5 to 8 per cent water
to develop adequate strength. They are less refractory than synthetic sands, as might be expected from the presence of impurities;
accordingly, their greatest use is for casting non-ferrous metals

and cast iron. *Synthetic sands* are washed and graded and contain no organic matter or clay. When they are bonded with the usual 3 to 5 per cent bentonite clay, moisture need not be higher than 3 to 4 per cent to develop good molding properties. Permeability is relatively high, and composition control is simple. Synthetic sands were originally exploited for steel foundry use because of their greater refractoriness as compared with natural sands, and because of their lower required moisture content. More recently, however, synthetic sands have become widely used for casting other types of metals, because of the close control of sand properties obtainable.

Sand-clay molding mixtures are used in the *green* (moistened) or *dried* conditions. *Green sand molds* are those in which the mold is closed and metal poured soon after molding, before appreciable drying has occurred. There may be from 3 to 8 per cent moisture (depending on the type of clay) in the sand in this condition. Green sand molds are used more than any other type because of the economy; no time is lost or expense incurred in driving off the water, since the mold is used *as is*. For most work this practice is satisfactory. Moisture can be controlled closely enough to prevent excessive steaming at the mold-metal interface, and permeability can be kept high enough to prevent *blows* caused by release of steam and other gases.

In *dry sand molds,* free moisture is completely removed by heating in an oven. Generally a harder, stronger mold results from drying, and less mold gases are present. Thus dry sand molds may produce more *dimensionally accurate* castings as compared to green sand molds, and are less susceptible to *breakage, cracking,* and *gas blows* than green sand molds. Some of the advantages of dry sand molds are obtained in *skin-dried* and *air-dried* molds. In skin-dried molds, only the surface moisture is evaporated by torch or warm air applied briefly to the mold cavity. Air-dried molds are molds that are allowed to stand in air for a considerable time before pouring.

2.4   Types and bonding action of clays

a. Types of clays

The three general types of clays for bonding molding sands are known industrially and scientifically as *montmorillonite* (sodium or

TABLE 2.1

COMPOSITIONS OF PURE BONDING CLAYS

$$Na_{0.33}$$
$$\uparrow$$
Sodium montmorillonite $(Al_{1.67}Mg_{0.33})Si_4O_{10}(OH)_2$
$$Ca_{0.33}$$
$$\uparrow$$
Calcium montmorillonite $(Al_{1.67}Mg_{0.33})Si_4O_{10}(OH)_2$
Illite $Al_4K_2(Si_6Al_2)O_{20}(OH)_4$
Kaolinite $Al_2(Si_2O_5)(OH)_4$

calcium *bentonite*), *kaolinite,* and *illite.* All three types occur naturally in extensive deposits which vary appreciably in purity; only relatively pure clays are suitable for bonding purposes. Clays are residual products of various kinds of basic rock. Montmorillonites are the weathered product of volcanic ash; illite is the decomposition product of micaceous materials; and kaolinite, the residue of weathered granite and basalt. The chemical composition of the active bonding minerals in these clays is indicated in Table 2.1.

The weathering action of atmosphere decomposes base rock into a fine crystalline powder; clays are generally ground before using to break up large pieces, which are actually densely packed aggregates of finely divided particles. Average clay particles are of colloidal size, as shown in Table 2.2.

b. Bonding action of clays

The science of clays and silica sands, particularly as applied to aggregates used for molding purposes, has not progressed to the point where we can state clearly what forces are involved in hold-

TABLE 2.2

AVERAGE PARTICLE SIZE OF VARIOUS BONDING CLAYS

Clay	Thickness in Millimicrons	Width in Millimicrons
Sodium bentonite	1	100–300
Calcium bentonite	1	100–300
Kaolinite	20	100–250
Illite	20	100–250

ing particles together. Accordingly, several theories have been advanced.

1. *Electrostatic Bonding of Clays*

Dry clay does not provide the necessary bond to hold sand grains firmly together; bond is developed only when clay particles are *hydrated*. In the presence of water molecules, which tend to hydrolyze according to the reaction

$$H_2O \leftrightharpoons H^+ + OH^-$$

clay particles preferentially adsorb hydroxyl (OH^-) ions, owing to unsatisfied valence bonds at the surface of the clay crystal, and the clay-water particle becomes negatively charged. As such, the clay-water *hull* attracts positive (H^+) ions in the surrounding water medium. The hydrogen *counterions* and the adsorbed hydroxyl ions about the clay particle comprise a so-called *double diffuse layer*. A hydrated clay particle or *micelle* is illustrated schematically in Fig. 2.2.

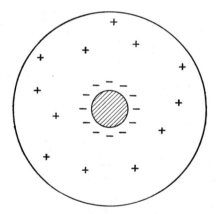

FIG. 2.2. Schematic representation of a clay micelle. Surrounding the clay particle are negatively charged hydroxyl ions positioned over a range of distances from the clay particle. Outside this layer, positively charged counterions (usually hydrogen ions) also are located at various distances from the clay center—hence the term double diffuse layer. This layer is rigidly attached to the surface of the clay particle and is considered to behave as a solid.

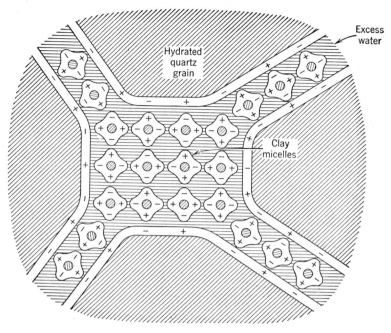

Fig. 2.3. Schematic sketch showing disposition of clay and quartz micellar dipoles. In green sand the intermicellar voids are filled with water.

Particles of sand (quartz) also form micelles by the adsorption of hydroxyl ions and hydrogen counterions. When quartz and clay micelles are formed in each other's presence, the hydroxyl ions of the clay micelle exhibit an attraction for the hydrogen counterions contained in the quartz micelles. The result is an electrostatic bond between sand and clay particles and between clay particles as sketched in Fig. 2.3.

2. Bonding by Surface Tension Forces

Forces developed by electrostatic interaction between sand-clay particles do not seem strong enough to account for all the strength properties of green sand mixtures—nor in particular for the high resistance to deformation of dry sand aggregates. Another possible source of bond strength is the surface tension of the water surrounding the clay and clay-sand particles, and filling the capillary interstices, particularly the interstices of the clay particles. This has been described by Norton,[1] and measured by Westman [2] for ball clays; values as high as 880 psi have been obtained, attributable to the surface layers of water acting on a stretched membrane, forcing the particles together. As the water layer becomes thinner by drying, the forces holding the particles together increase.

3. *Bonding Due to Interparticle Friction*

Still another force which can add to the strength of clay-sand mixtures is that due to the geometry of the aggregate. The so-called "block and wedge" theory proposed by Grim [3,4] involves essentially the interparticle friction developed in non-plastic particulate materials under pressure. Since sand or sand-clay particles are packed inside a flask, most particles are jammed against their neighbors. The resultant interparticle friction opposes further deformation, and causes a bridging action between long rows of favorably oriented particles and the sides of the flask. That this is a source of some bonding force is indicated by the sharp change in strengths of sand mixtures which results from blending grain sizes and shapes favorable to the development of interlocking, frictional forces.

2.5   Sand additives

Although the basic ingredients of a sand "mix" are only sand, clay, and water, other materials are often added in small amounts for special purposes. So many materials (additives) are used for so many different purposes that it is possible here to present only a few of the most important.

One common reason for adding such materials to sand molds is to minimize *sand expansion* defects. Figure 2.4 shows the expansion characteristics of a silica (quartz) sand between 0°F and 1800°F (982°C). At 1063°F (573°C), a rapid increase in volume occurs owing to the transformation of "low" quartz to "high" quartz. At higher temperatures, still other transformations occur with corresponding volume changes. The transformation at 1063°F (573°C) probably accounts for most *sand expansion* defects, especially in non-ferrous metals. Such defects result when the interface sand, heated by radiation or conduction from the metal rising in the mold, increases in volume and shears away

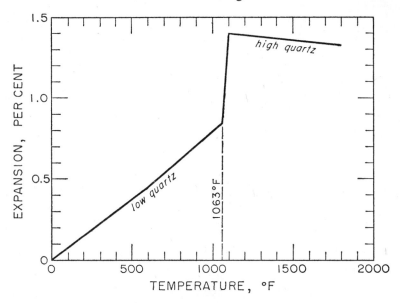

Fig. 2.4. Thermal expansion of silica. After Fairfield.[5]

from deeper adjacent layers which have not been heated. Mold *washes, scabs,* or *buckles* may result. "Buffer" materials such as wood flour, fine sawdust, cereal, grain hulls, or perlite are added to the sand to eliminate such defects.

Sand additives are sometimes used to alter various mechanical properties of sand molds. *Green strength* and *dry strength* of a sand-clay mixture may be increased by the addition of one or more of the various cereals. Dextrin is used to increase *dry strength*. *Hot strength* is increased by silica flour, iron oxide, or ground pitch. *Collapsibility* is improved by cereals, or by other organic bonding agents, which "burn out" when heated to high temperatures by the molten metal in the mold.

Additions are also made to sand molds to improve the surface smoothness of castings. Carbonaceous materials (ground pitch, sea coal, and gilsonite) are used to obtain *fine surface finishes* in ferrous castings. Boric acid and sulfur are used in molds for magnesium casting, to prevent *metal-mold reactions;* diethylene glycol or cereal may be added to bentonite bonded sand molds to reduce their tendency to *surface drying* and thereby improve casting surface finish. Many proprietary materials are sold to the foundry industry by supply houses as mold additives for various purposes.

2.6   Preparation and use of molding sands

Sands may be prepared for molding by mixing the ingredients (sand, clay, cereal, water, and such other materials as may be required or desired) by hand shovel, or with one of a variety of mechanical mixers. To develop optimum properties in molding sand, it is necessary that each grain be coated uniformly with clay and that the clay be *tempered* (uniformly wetted) with water. A definite *mulling* action is required for thorough mixing, in which sand grains, clay, and water are rubbed intimately together. Hand mixing does not accomplish this, nor do some of the simpler, cheaper mechanical mixers sold for the purpose; such methods have the advantage of economy in first cost and maintenance, which appeals to many foundrymen, but may ultimately be more expensive in terms of scrapped castings. *Sand kickers* are carriages equipped with mechanically driven blades which straddle long heaps of sand on the foundry floor; water, poured on the heaps previously, is cut into the sand as the kicker moves along the heap. Screenerators are units equipped with vibrating screens through which sand is shifted before being thrown from a rapidly rotating drum or belt into heaps or bins. Some mullers depend upon the weight of huge wheels rolling relatively slowly over the ingredients held in a stationary pan; cutter plows revolve with the rotating assembly to keep sand stirred beneath the wheels. In some units the pan revolves slowly, and plow and wheels are fixed in position. In other units wheels rotate rapidly around a vertical axis inside a rubber-lined cylindrical housing. The ingredients are kept in constant agitation and are mixed thoroughly under the centrifugal force of the small wheels revolving at high speed. Preparation and handling of molding sands are discussed in greater detail in Chapter 11; Figs. 11.1 to 11.7 illustrate the various sand mixing units mentioned above.

2.7   Cores

The use of cores to obtain internal configurations has been described and is sketched in Fig. 2.1. Figure 2.5 illustrates the steps involved in making a typical core. Cores are usually made

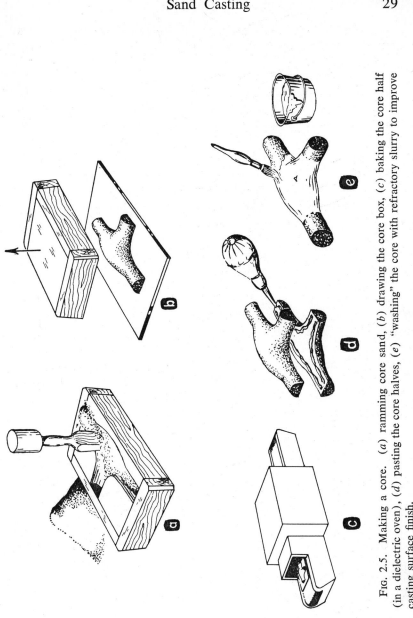

Fig. 2.5. Making a core. (a) ramming core sand, (b) drawing the core box, (c) baking the core half (in a dielectric oven), (d) pasting the core halves, (e) "washing" the core with refractory slurry to improve casting surface finish.

of synthetic sand, although clean, natural sand containing only 1 or 2 per cent clay can be used. Green sand cores are employed occasionally, in which event a strong, highly permeable molding sand is used. Cores may also be made of green sand used in the dried condition. Most frequently, however, they are bonded with an organic agent such as linseed oil; cereals may be added to make the raw mixture stronger. The basic advantage of organic binders for cores (as compared to clays) is that they break down under the heat of the metal (have *collapsibility*) and so can be easily removed from the casting at shakeout.

Cores are made by *ramming* or *blowing* the raw sand mixture into coreboxes. Metal rods are sometimes used internally to make the core stronger. If cores are made in parts, these are pasted together after baking and are rebaked briefly to dry and set the paste. Baking is done in an oven, preferably with circulating air, at about 450°F (230°C) until the core reaches nut-brown color. Oxidation, condensation, polymerization, and simple drying develops strength in the core according to its original composition. It is extremely important that proper baking times and temperatures be established for the various binders and for variations in core size. A properly baked core does not produce harmful gases, has adequate strength, and collapses at the right time after metal is poured around it.

Recently, plastics have begun to supplement linseed oil as core binders. Urea formaldehyde and phenol formaldehyde are the two most widely used. Urea formaldehyde breaks down at very low temperatures as compared to phenol or linseed oil, and so its use is desirable in light (thin) section castings of the lower-melting metals. Formaldehydes have the advantage that they can be baked extremely rapidly by high-frequency electromagnetic waves, i.e., in *dielectric ovens*.

A still more recent development in core making is commonly termed the "CO_2 process." In this process, sodium silicate (Na_2SiO_3) is used as the binder, generally with organic additions to improve collapsibility. After the core is rammed but *before* it is removed from the corebox, CO_2 gas is passed into the core, generally through a series of vents. The sodium silicate then *sets* by the reaction

$$Na_2SiO_3 + CO_2 \rightarrow Na_2CO_3 + SiO_2 \quad \text{(colloidal)}$$

Advantages of the CO_2 process are that it does not require expensive baking, and the core is fully hardened before it is removed from the corebox. Breakage and sagging of unhardened cores which often occurs with other binders is eliminated. Expensive *driers* to support intricate cores during the baking operation are unnecessary.

2.8   Molding methods

a.   Flask molding

The basic method of molding, hand ramming, has been described. The chief advantage of hand molding is no expenditure for equipment. Certain intricate and artistic work cannot be done

Methods of Ramming

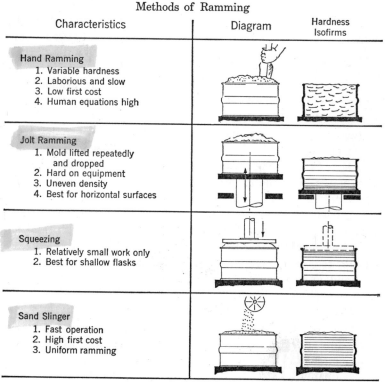

Characteristics	Diagram	Hardness Isofirms
Hand Ramming 1. Variable hardness 2. Laborious and slow 3. Low first cost 4. Human equations high		
Jolt Ramming 1. Mold lifted repeatedly and dropped 2. Hard on equipment 3. Uneven density 4. Best for horizontal surfaces		
Squeezing 1. Relatively small work only 2. Best for shallow flasks		
Sand Slinger 1. Fast operation 2. High first cost 3. Uniform ramming		

After Buchanan

FIG. 2.6.

by machine molding on a production basis; nor does it pay to adapt patterns to molding machines when only a few castings are to be made. Characteristics of various molding methods, discussed below, are given in Fig. 2.6. *Isofirms,* or lines of roughly equal mold hardness, are shown as a schematic means for differentiating practically between the various methods. Actually flow lines are much more complex and variable than shown.

In *squeezer* and *jolt* ramming the flask is filled with sand by shovel or overhead hopper. In *squeezer* molding, a platen slightly smaller than the inside dimensions of the flask is brought into contact with the upper surface of loose sand and pressure is applied steadily through an air valve actuated by the molder's knee or hand. The platen may move downward, or may be stationary with the mold moving upward. In the *jolt-squeeze-draw* unit, a vibrator facilitates removal of the pattern as the mold base or platen returns to molding position. In *jolt* molding, pressure is applied suddenly by raising the mold a short distance and allowing it to drop against a solid bed plate; this pile-driver action is continued until adequate mold hardness is attained. *Jolt-squeeze, jolt-squeeze-rollover,* and other combinations of sand-packing machinery are available; the roll-over units clamp and turn the drag automatically to facilitate ramming the cope or drawing the pattern.

Figure 2.7 illustrates the operation of one type of molding machine (jolt-squeeze-rollover). The pattern, mounted on two sides of a board (or matchboard) is placed between the empty cope and drag flasks, with the drag side up. The drag half of the flask is filled with sand, and the assembly jolted (Fig. 2.7*a*). After strike-off, a squeezer platen is placed over the drag and the assembly inverted (Fig. 2.7*b*). The cope half is then filled with sand and both halves are squeezed simultaneously (Fig. 2.7*c*). Finally, the cope is removed automatically, the pattern "drawn" from the drag by hand, and any necessary finishing or core setting is done before reclosing (Figs. 2.7*d, e*).

The sand slinger is an automatic unit which effectively *throws* handfuls of sand into the flask with machine-gun rapidity and great force. High uniform mold hardness can be developed by this method (Fig. 2.6). Figure 2.8 illustrates one type of relatively small sand slinger. Some sand slingers are sufficiently large so that a seat is provided for the operator over the position of sand ejection. Power-driven controls enable him to pilot the head of

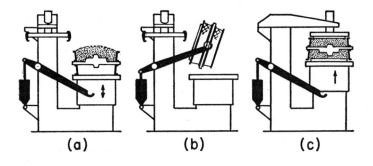

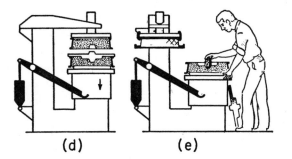

FIG. 2.7. Operation of a jolt-squeeze-rollover molding machine.

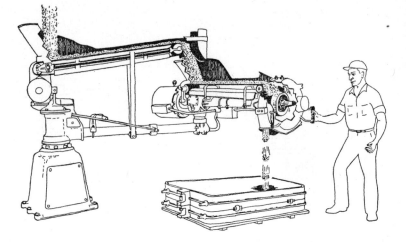

Fig. 2.8. Sand slinger.

the sand slinger back and forth over large molds, or from mold to mold.

b. Loam molding

Loam molding was much used in the past for making large bronze castings and is still practiced in some shops, particularly for making huge manganese-bronze propellers. In this method a substructure is made of bricks, wood, and other material to the approximate contour of the casting. A very viscous slurry of water, clay, and sand is daubed over the framework and worked to proper shape with *sweeps*. The mold is dried by forced hot air or torches. No pattern is required, as sheet-steel sweeps are so shaped that they generate proper casting contour as the sweep arm is moved back and forth over a fixed spindle. Such sweeps are used occasionally in making molds for large rolling-mill rolls where ordinary molding sand is used instead of a slurry. The chief advantages of this process are savings in pattern cost and storage; pattern storage alone is an important and expensive item in most foundries. Loam molding is slow and laborious, and special molders are required; all the work must be done by hand as the process is very much an art. Figure 2.9 illustrates the loam molding process as applied to the manufacture of a large bell.

c. All-core molding

Some large castings are made entirely of cores; hundreds of large and small individual core sections are fitted together in a pit and finally rammed securely in position with molding sand or pressed together with suitable clamps. Some large cores for this purpose weigh as much as 60 tons. This method is chosen where design is such that standard patterns could not be drawn from regular molds. Raw core sand is sometimes rammed into flasks and used as regular molding sand; the mold is then baked like a core. This method is chosen when (1) a casting of extremely accurate dimensions is required, or (2) many cores must be carefully set; also, core sand molds are sometimes used to improve casting surface finish, or permit casting very thin sections. Cores are often used for *insert molding* where special metal shapes are inserted in a

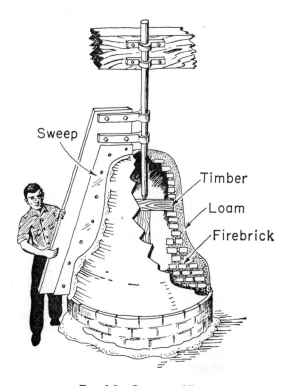

Fig. 2.9. Loam molding.

mold and metal is cast around them. For example, preformed
turbine blades may be held firmly in core sand with part of each
blade protruding into the cavity where the rotor is to be cast; after
casting, the blades and rotor are an integral part. Special precautions can be taken to get a metallurgical bond (welding) between
the inserts and the casting, or a simple mechanical interlocking is
sometimes satisfactory. Cores are usually used in this type of
molding because greater accuracy of alignment can be obtained
with cores than with green sand molds, and there is less chance for
shifting of the inserts during molding.

d. Shell molding

So-called *shell molding* is a special form of sand casting; the
process is a relatively recent development especially adapted to
high production. Details are shown in Fig. 2.10. Skilled molders
are not required, and the process can be highly mechanized.

Sands for shell molding are always washed and graded for best
results; fine sands can be used, as permeability of the thin shell is
not a problem. Thermosetting plastics are currently used for bonding the sand grains; dry powder and sand are mixed intimately in
a muller or other suitable device, or sand may be purchased "precoated" with resin. Metal patterns sprayed with a separating
silicone grease are used for shell molding; they are heated to about
400–450°F (205–230°C) and covered quickly with the resin-
bonded sand. In about 30 seconds a hard layer, or shell, of sand
forms over the pattern; pattern and shell are usually heat-treated
in an oven for about 2 minutes at about 600°F (315°C), and the
shell is ready to strip from the pattern. A complete mold is made,
in two or more pieces; cores may also be made by this technique.
The shells are clamped securely and usually embedded in gravel,
coarse sand, or metal shot; the mold is then ready for pouring.

Highly specialized equipment is available for this process or
can be designed and made independently. Castings can be made
of any metal, and excellent surface finish and good dimensional
control are possible. The future for this new process is very
bright.

2.9   Sand control

Test methods available to the foundryman for controlling the
properties of his molding sands, although more or less accepted as
standard, do not measure any unique constant of the material in

a physical sense, but they do furnish useful, practical indication of molding sand performance. They are described in detail in "Standards and Tentative Standards for Foundry Sand Control," a publication of the American Foundrymen's Society.

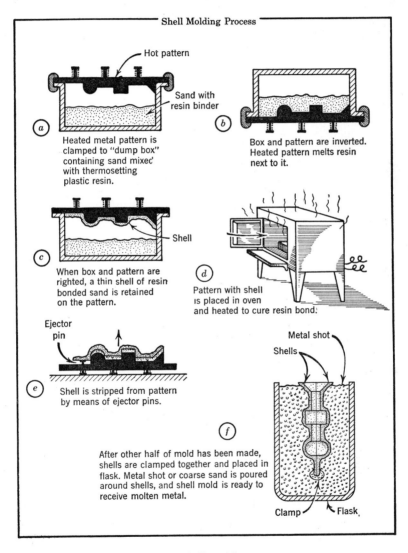

Shell Molding Process

(a) Heated metal pattern is clamped to "dump box" containing sand mixed with thermosetting plastic resin.

(b) Box and pattern are inverted. Heated pattern melts resin next to it.

(c) When box and pattern are righted, a thin shell of resin bonded sand is retained on the pattern.

(d) Pattern with shell is placed in oven and heated to cure resin bond.

(e) Shell is stripped from pattern by means of ejector pins.

(f) After other half of mold has been made, shells are clamped together and placed in flask. Metal shot or coarse sand is poured around shells, and shell mold is ready to receive molten metal.

Fig. 2.10. Shell molding process.

Some of the more universally used tests are designed to measure the following properties: (1) *permeability,* the capacity of a sand to vent gases away from the mold cavity; (2) *hardness,* the resistance of the mold surface to deformation by a standard indentor; (3) *moisture;* (4) *shear* and *compressive* strength, measured on cores or green and dry specimens of molding sand by standard apparatus on standard specimens; (5) *hot strength,* measured by compression of a standard test piece in a furnace capable of reaching 3000°F (1650°C); (6) *collapsibility,* indicated by failure of the specimen at high temperature under load; (7) *expansion* and *contraction* characteristics, checked by special attachments to the same furnace; and (8) *grain size, shape,* and *distribution,* determined by the usual methods for particle classification or petrographic analysis. All these are described in detail in the reference mentioned.

2.10   Grading of sands

Until fairly recently the standard classification for foundry sands was the A.F.S. fineness number also described in the reference given; it is an arbitrary value which gives no indication of grain size or distribution, although analysis is made on Tyler or U. S. Standard sieves. Unfortunately, sands are still widely specified in this way, but they can be defined much more uniquely by plotting the data obtained from standard sieves on a cumulative curve. This is simply a plot of cumulative per cent of sand retained on successive sieves against the sieve number, or particle size in microns. Three-cycle, semi-log paper is used to condense the curve. Clay and other *fines* in natural sands are separated by washing the sand before sieve analysis, and this fraction is analyzed by hydrometer or elutriation methods to complete the cumulative curve beyond the limits of the finest sieve. A typical curve is shown in Fig. 2.11. The *coefficients of skewness* and of *sorting,* or the entire cumulative curve, may be used to specify sands so closely that their performance will be dependable from day to day. The serious limitation of the A.F.S. number is best illustrated by the fact that two sands, with the same fineness number, can be respectively a synthetic sand suitable only for steel casting and a natural sand for non-ferrous use.

Not all sands are suitable for molding purposes, even though their grain shape and distribution are satisfactory. In order to

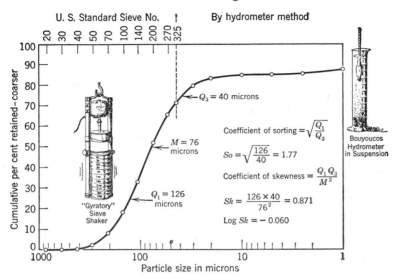

FIG. 2.11. Cumulative curve for A.F.S. 110 naturally bonded sand.

define sands completely so that various shipments will give rigorously similar results, chemical composition, and grain structure must be checked. Chemical impurities, feldspar for example, and inorganic impurities in natural sands, are deleterious if present in too great proportion. Grain shapes vary from sharply angular, through subangular, to rounded; some foundrymen have preferences for certain grain shapes, but no quantitative data exist on the relative performances of the various types. As sands tend to segregate during shipping, storing, and handling, care must be taken that they are well mixed before sampling or using.

2.11 Molding materials other than silica sand

The relatively low cost, and adequate refractoriness (for most purposes), of silica sand has made it the most widely used molding medium in the foundry industry. Other refractories, however, can be obtained in fine particle sizes and molded with suitable binders in the same manner as silica sand. Some properties of selected refractory materials are listed in Table 2.3; many of these materials are used for special purposes, i.e., mold inserts, in the sand casting process.

TABLE 2.3

PROPERTIES OF SELECTED REFRACTORY MATERIALS *

| | Melting Point | | Linear Coefficient of Expansion, 80°F–1470°F, |
Refractory	(°F)	(°C)	(× 10^6/°C)
Silica (SiO_2) †	3115	1710	16.2
Alumina (Al_2O_3)	3670	2020	8.0
Beryllia (BeO)	4660	2570	7.5
Magnesia (MgO)	5070	2800	13.5
Thoria (ThO_2)	5530	3050	9.5
Zirconia (ZrO_2)	4890	2700	6.5
Zircon ($ZrO_2 \cdot SiO_2$)	4800	2650	4.5
Spinel ($Al_2O_3 \cdot MgO$)	3870	2130	8.5
Mullite ($3Al_2O_3 \cdot 2SiO_2$)	3290	1810	5.5
Sillimanite ($Al_2O_3 \cdot SiO_2$)	3290	1810	—
Silicon carbide (SiC)	Decomposes before melting 4000–4900°F (2200–2700°C)		3.5
Carbon baked (C)	Sublimes 5800–7600°F (3200–4200°C)		0.65
Carbon graphite (C)	Sublimes 5800–7600°F (3200–4200°C)		—

* Data (except for silica) from Smithells.[15]
† From Norton.[1]

In steel foundries, zircon sand ($ZrO_2 \cdot SiO_2$) is used as a *moldable chill* since it is able to extract heat from a casting somewhat more rapidly than silica sand. Additional advantages of zircon sand as compared to silica are (1) it possesses a small expansion coefficient and no phase transformation, and so is less liable to produce expansion defects, (2) its high fusion point prevents possible "burn in" or metal penetration at heavy metal sections, and (3) it presents no *silicosis* health hazards as does fine silica when inadequate safety precautions are observed. *Olivine sand* [(Mg, Fe)$_2$ SiO$_4$] is used for much the same reasons, especially in Europe. *Chamotte* (calcined fire clay) is also occasionally used as a facing material to prevent expansion and penetration defects in ferrous castings. Large castings made in chamotte have exceptional external appearance and do not show *scabs* and *sand spots* as do ordinary sand castings. Chamotte is outstanding in its resistance to erosion by molten steel and is useful for runner blocks and cores as well as molds.

Some of the refractories listed in Table 2.3, including carbon, thoria, beryllia, and magnesia, are used in rammed refractory molds for casting the very reactive metals: titanium, zirconium and uranium. These and other refractories are also used in the *investment casting* process to be discussed in the following chapter.

2.12 Melting metals for casting

Manufacture of castings is essentially a matter of heat transfer in one or another form. Heat is first added to the cold, solid metal (scrap or ingot) for melting and for superheating the molten metal until it is fluid enough to pour into a mold. Various types of furnaces (described in detail in Chapter 12) are used for this purpose. Heat is then extracted from the metal by the mold to re-form it into a solid, cold body of desirable size and shape. How the heat is added and removed affects the quality, economy, and other characteristics of the product.

All types of metal-melting furnaces can be used for foundry operations; however, foundry requirements are sometimes unique and one or another type of furnace may be best for a particular operation. For example, open-hearth furnaces are desirable in large foundries requiring 10 to 200 tons of molten metal at one time. In a smaller jobbing shop, where 1 to 15 tons of metal may be needed at frequent intervals, an electric-arc furnace may prove best. In still another foundry requiring only 250 lb to 1 ton of alloys of various special compositions, an induction furnace will give best all-around performance. The choice of furnace may be dictated by (1) considerations of initial cost, (2) relative average cost of maintenance and repair, (3) base cost of operation, (4) availability and relative cost of various fuels in the particular locality, (5) cleanliness and noise level in operation, (6) melting efficiency, in particular the speed of melting, (7) degree of control (metal purification or refining) required, (8) composition and melting temperature of the metal (for example, a cupola is normally used for melting cast iron but seldom, if ever, for melting aluminum or magnesium), and (9) personal choice or sales influence.

Fuels or power sources for melting metals, and the types of furnace with which each is normally used, are listed in Table 12.1. Various types of foundry melting furnaces are sketched in Figs. 12.1 to 12.8.

2.13   Pouring practices

Metal is seldom poured into molds directly from the melting furnace; some form of *ladle* (Fig. 2.12) is required for pouring control and increased maneuverability.   Sometimes metal is poured from one or more furnaces into a large *holding ladle,* of many tons'

Fig. 2.12.

capacity, later poured into smaller *bull ladles* which can be moved by monorail to the pouring flour, and then distributed to the molds by even smaller *hand* or crane ladles. *Hand ladles, two-man ladles,* and *bull ladles* hold about 60, 180, and over 200 lb of metal respectively. Bull ladles may hold 100 tons or more. Pouring ladles are handled manually or by monorail or overhead crane, with special pouring fixtures.

Slag and impurities which collect on metal during melting and pouring must be kept from entering the mold. This is sometimes difficult with simple *lip pouring ladles,* from which metal is poured into the mold by tilting the ladle and allowing the metal to run over the upper rim. *Teapot-pour ladles* are equipped with a pouring spout extending down their entire depth, with an opening at the bottom; in this way only clean, slag-free metal is poured. *Bottom pouring* is used particularly for iron and steel, with ladles from 2 to 100 tons' capacity. In this type a hole from 1 to 4 in. in diameter is formed in a refractory brick, which is molded into the ladle bottom. The hole is opened or closed by a tile-covered *stopper rod,* with a blunt, round, refractory *point* on the end, extending down through the molten metal, and fastened securely to an arm actuated by a lever at one side of the ladle.

2.14   Cleaning, inspection, and repair

Castings are removed from the mold by hand or mechanical shaker. In the hand method molds are inverted and sand and castings shaken into a pile from which the castings are retrieved and placed in a *tote box;* the sand is cooled and *reconditioned* for use. Mechanical methods are always used for production work; *vibrating shakeout units* are simply mechanical vibrators upon which flask and casting are placed; the sand falls through grates, and casting and flask work their way forward to where they fall into pans or are picked up. In the *hydroblast unit,* to be described in detail later, sand is removed from flask and casting by high-velocity streams of water. After *shakeout,* the sand is either reclaimed or dried and reused directly.

Gates and risers are removed from castings by one or another of several processes. *Flame cutting* is used mainly for removing gates and risers from steel castings; only recently have methods been developed for its use with non-oxidizable metals. Fins, pads, or other unwanted protuberances can be removed by *torch, chip-*

ping hammer, or *saw;* all rough places are smoothed with a *grinding wheel.* Final cleaning operations include *sand* or *shot blasting, pickling,* or *tumbling.* In tumbling, castings are placed in a rotating steel drum with small, sharp-pointed objects called *mill stars.* The tumbling and abrading action, which may be by wet or dry method, cleans the castings and will also break gates and risers from small iron castings. Sand blasting and tumbling are employed simultaneously in some mechanical units.

Unfortunately, most castings receive a superficial inspection by eye only. There are many critical methods of inspection such as magnetic powder, fluorescent powder, X-ray, and ultrasonic reflectoscope, which can be used to insure quality castings; such methods are enjoying increased popularity with producer as well as consumer and will be described in detail later. It is extremely important that unsatisfactory castings be found at the foundry rather than later, after considerable machining and other work have been done on them. Determining quality at this point also encourages immediate improvements in practices, if necessary.

Defects in castings, if not too severe, may be repaired by *welding, brazing,* or *soldering.* *Arc* and *gas welding* are standard practices for repair of steel castings and are being used more and more for iron and non-ferrous metals. *Burning-in* is sometimes done by packing molding sand around a defective area and flowing molten metal from a ladle over the defect until the casting has melted locally and fusion is obtained. Exothermic materials which react to produce metal at 4000–5000°F (2200–2800°C) are also used for repairing defects in alloys of iron, copper, and nickel.

Aluminum, magnesium, brass, bronze, and iron castings are often porous as the result of interdendritic shrinkage; such defects can be closed by impregnation with sodium silicate, Bakelite, resins, and other materials forced into the interstices under pressure. Such materials harden and seal pores against leakage. Surfaces of castings may also be *wiped* with solder to close voids and prevent leakage. The method of casting repair varies with metal, defect, and service requirements. Bronze castings are sometimes heated to oxidize small pores and so prevent leakage. *But all these methods are poor substitutes for sound castings.*

2.15   Summary

This chapter introduced the casting as a dependable engineering material and in particular the process of sand casting by which it is made. We learned that it is necessary to consider economics and the mechanical properties required before determining whether a part is to be made by casting, welding, or metal working; and we also learned that castings have certain unique properties.

Sand casting is popular because it is a versatile, high-production process, amenable to manufacture of castings from nearly any metal that can be melted; titanium casting will probably be an important exception. Sand is a relatively cheap refractory, available in unlimited quantities, and extremely adaptable to molding processes. The basic principles of molding and various molding methods were described.

Clay is used for bonding molding sand, and linseed oil, resins, and plastics are used for making cores. Illite and kaolinite clays (the fireclays) are decomposition products of rock, and montmorillonite (bentonite clay) comes from chemical and mechanical weathering of lava. The bonding action of clay on sand is electrostatic in nature and not a simple chewing-gum phenomenon as might be thought. Baking cores consists of drying, condensation, polymerization, or oxidation, depending upon the type of binder used.

Patterns and pattern making are not discussed in detail except to emphasize that patterns should be durable and resistant to moisture and warping, and should have excellent surface finish; further that patterns should be made only after due consultation among designer, pattern maker, and foundryman. The time for agreement is before, not after, design is determined. It can rightly be said: "As the pattern goes, so goes the casting."

Methods of making flask-type molds include *hand ramming, jolt ramming, squeezing, jolt-squeezing,* and use of a *sand slinger.* These methods were discussed from the standpoint of their effect on the mold, and typical molding equipment was illustrated. Other methods of making sand molds were described; these include *loam molding, all-core molding,* and *shell molding.*

Shop control of sand involves testing for the following properties: permeability, hardness, moisture, and green and dry compressive strength; more complete testing for research and development work also includes hot strength, collapsibility, expansion and contraction characteristics, and measurements of grain size, shape, and distribution. The cumulative curve is much more desirable than the A.F.S. fineness number for specifying molding sands.

Sometimes sands are used repeatedly without reclaiming, in

which event it is difficult, if not impossible, to obtain uniform properties. Reclaiming may be done by wet or dry methods, and it is always desirable (from the standpoint of casting quality) unless a new sand facing is used next to the pattern. Besides improving casting quality, it may prove economical to reclaim sand, particularly if freight charges are excessive. Used sand can be restored to original quality, and even original color, by proper cleansing, filtration, and heat treatment. Sand reclamation is discussed in Chapter 11.

Pouring practice involves use of lip-pouring, teapot-pouring, and bottom-pouring ladles, or pouring into molds directly from the melting unit. Cleaning, inspection, and repair of castings were considered briefly; they will be discussed in more detail in Chapter 9.

All types of melting units are used in foundries. In general, cast iron is melted mostly in cupolas, with the arc and open-hearth units increasing slowly in popularity; air furnaces, a special form of the open hearth, are used extensively in shops making malleable iron. Steel for castings is commonly melted in a three-phase electric-arc furnace of size suitable for the particular operation; open-hearth furnaces are used in large foundries and ingot shops, as are cupola-converter combinations. Aluminum and copper-base alloys are melted in various types of crucible furnaces, indirect-arc furnaces, and small open-hearth units; magnesium alloys are usually melted in iron crucibles set in oil- or gas-fired furnaces. The induction furnace can melt nearly all commercial metals and alloys. The choice of melting unit may be based on one or more of nine influences mentioned in this chapter. Melting practices for cast metal are treated in detail in Chapter 12.

2.16 Problems

1. List four economic or metallurgical reasons for fabricating metal parts by casting.

2. List the advantages of natural molding sand; of synthetic molding sand.

3. List the principal parts of a mold, and describe their functions.

4. Why is backing sand used instead of using all facing sand?

5. List the advantages of a matchplate type of pattern.

6. Why are patterns generally made oversize?

7. List the various molding materials used in the foundry. Indicate the alloys cast therein.

8. Differentiate critically between green sand, dry sand, and skin-dried molds.

9. What bonding agents are used in molding and core sands?

10. When are chamotte molds preferable to sand molds? For what metals are chamotte molds particularly suitable?

11. (a) List the requirements for developing bonding strength in clay-bearing molding sands. Can alcohol be used to plasticize clay? Explain. (b) List three basic clay types. What is the genesis of each?

12. List four chemical or physical processes that may occur in baking cores.

13. Draw isofirms to indicate the densification of molding sand by squeezing, jolting, and sand-slinger processes; also those produced by hand ramming.

14. List advantages for loam molding and all-core molding methods.

15. Why are the following materials added to sand molds: (a) wood flour, (b) pitch, (c) boric acid, (d) diethylene glycol?

16. What are the advantages of shell molding as compared to green sand molding? What are the disadvantages?

17. Why are mold materials other than sand and clay sometimes used for metal casting?

18. Why are teapot- and bottom-pouring ladles used? Would you use them for casting magnesium?

19. List the cleaning operations you would be likely to use for a small steel casting.

20. Prepare small, simple patterns (flat shapes are adequate) of the materials listed below, and prepare a mold with all of them in the same flask; then remove the patterns, and observe the smoothness and sharpness of detail and any other features of the mold cavity that might be influenced by the material from which the pattern is made. This should impress upon your mind the desirability of using the very best material for patterns and of preparing the pattern carefully.

(a) Soft wood, sanded smoothly and coated with 1 coat of shellac.
(b) Soft wood, sanded smoothly and coated with 3 coats of shellac.
(c) Hard wood, sanded smoothly and coated with 3 coats of shellac.
(d) Polystyrene.
(e) Steel.
(f) Aluminum.

21. Mix samples of (a) natural green sand and (b) synthetic green sand bonded with bentonite, and pack a quart mason jar tightly with each; then (a) place a drop of water on each, leave for 2 minutes, and observe any differences in water migration that may exist, and (b) screw the jar tops on tightly and let stand overnight. These experiments should illustrate a difference in the nature and rate of water

absorption which you will be able to explain if you investigate the structural differences between kaolinite and montmorillinite minerals.

22. Under average foundry conditions cores bonded with linseed oil are "baked" from 2 to 4 hours at about 300 to 450°F. Take two identical cores of about 3 in. thickness, and (a) heat one in an oven in the usual manner, and (b) blow air at the same temperature through the other. Note the difference in "baking time." What do you think would happen to the baking time if you added some oxygen to the air blown through the core? Try it.

23. Using a foundry riddle (sieve) build up layers of loose sand (each layer being 1 in. thick) in suitable containers (flasks), using carbon black between each layer. Consolidate the mass by (a) ramming, (b) jolting, and (c) vibrating; then slice through the mold vertically at various points, and observe the behavior of the isofirms which are delineated by the black streaks.

24. Use water to simulate metal, and pour into glass containers of various heights, shapes, and sizes; observe the flow characteristics and the tendency of the more turbulent conditions to cause gas bubbles to be stirred into the water. Do you think this tool could be used to predict the flow characteristics of molten metal in sand molds?

25. Write to a supplier of foundry sands, and obtain sieve analyses of their various grades of sand. Plot these on a cumulative curve, and note the differences between sands intended for use with different metals.

2.17 Reference reading

1. Norton, F. H., *Refractories,* McGraw-Hill, New York, 1949.

2. Westman, H. E. R., "The Capillary Suction of Some Ceramic Materials," *Journal American Ceramic Society,* **12,** 585, 1929.

3. Grim, R. E., and Cuthbert, F. L., "The Bonding Action of Clays, I, Clays in Green Molding Sands," *Illinois State Geological Survey Report of Investigations,* **102,** 1945.

4. Grim, R. E., and Cuthbert, F. L., "The Bonding Action of Clays, II, Clays in Dry Molding Sands," *Illinois State Geological Survey Report of Investigations,* **110,** 1946.

5. Fairfield, H. H., "Expansion of Silica Sands," *Foundry,* **76,** 128, May 1948.

6. Hauser, E. A., *Colloidal Phenomena,* McGraw-Hill, New York, 1939.

7. *Steel Castings Handbook,* S.F.S.A., Cleveland, 1950.

8. *Cast Metals Handbook,* fourth edition, A.F.S., Des Plaines, Ill., 1957.

9. Briggs, C. W., *The Metallurgy of Steel Castings,* McGraw-Hill, New York, 1946.

10. *Foundry Sand Handbook,* A.F.S., Des Plaines, Ill., 1953.

11. Dietert, H. W., *Foundry Core Practice,* A.F.S., Des Plaines, Ill., 1950.

12. *Symposium on Molding Machines,* A.F.S., Des Plaines, Ill.

13. Heine, R. W., and Rosenthal, P. C., *Principles of Metal Casting,* McGraw-Hill, New York, 1955.

14. *Foundry Practice,* U. S. Navy Bureau of Ships, Washington, D. C., U. S. Government Printing Office, 1944.

15. Smithells, C. J., *Metals Reference Book,* Interscience, New York, 1955.

CHAPTER 3

Casting Processes Other Than Sand Casting

3.1 Introduction

A casting process, for commercial success, may depend upon speed of production, improved smoothness of casting surface and/or dimensional accuracy, or upon some particular feature of special interest to the arts or professions. For example, *die* and *permanent mold* casting are appealing because of high production rates and because the molds in which castings are formed are not expendable; some molds are used for several thousand castings.

Die castings can be made in extremely thin sections and close enough to finished dimensions for assembly with little or no machining. Some alloys cast in dies and permanent molds yield castings slightly stronger and more ductile than sand castings because faster cooling rates impart a denser, finer-grained structure. *Investment casting,* popularly called *precision casting* and *lost wax casting,* enjoys commercial success because castings such as turbine blades for jet engines may be made to dimensional accuracy approaching 0.005 in. per in., and castings can be made of alloys which can be neither machined nor otherwise worked to shape. Investment casting is also useful to *dentists* for making dentures, to the *medical profession* (orthopedic surgery in particular) for implants, splints, and screws, to jewelers for costume and special jewelry, and lastly to sculptors for making statuary and other *objets d'art*.

3.2 Plaster casting

Plaster of paris (gypsum, or $CaSO_4 \cdot nH_2O$) is used for casting silver, gold, aluminum, magnesium, copper, and alloys of these metals, particularly brass and bronze (Fig. 3.1). Gypsum may be used as an *investment* casting, or for *cope and drag* molding.

50

The term *plaster casting* is reserved for the latter application, in which copes and drags are made and a *parting line is established* as in ordinary sand molding. To prepare plaster for use, *100 parts* of a special mixture purchased as *metal casting plaster* is

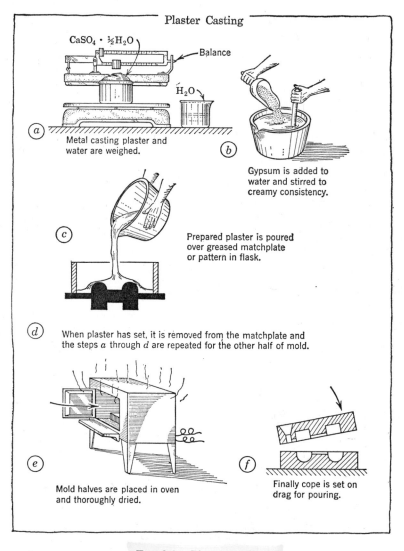

Plaster Casting

$CaSO_4 \cdot \frac{1}{2}H_2O$

Balance

H_2O

(a) Metal casting plaster and water are weighed.

(b) Gypsum is added to water and stirred to creamy consistency.

(c) Prepared plaster is poured over greased matchplate or pattern in flask.

(d) When plaster has set, it is removed from the matchplate and the steps *a* through *d* are repeated for the other half of mold.

(e) Mold halves are placed in oven and thoroughly dried.

(f) Finally cope is set on drag for pouring.

FIG. 3.1.   Plaster casting.

added to *160 parts* water and stirred slowly to creamy consistency; it is important that the plaster be added to water rather than water to plaster, and that an optimum rate of mixing be developed by experience for the particular mixer used; very rapid mixing develops too much air in the slurry, which adversely affects mold texture, and mixing too slowly may permit the slurry to harden (set) prematurely. The slurry is poured over a carefully made matchplate-type pattern, usually of metal; in a few minutes at room temperature the mixture develops an initial set, and the pattern can be removed (sometimes this period is hastened by heating or by adding a small quantity of terra alba). Copes and drags may be made simultaneously on separate lines and dried in ovens held at 400–800°F (200–425°C) until all free and combined moisture is removed. Mold sections are very fragile and require care in assembling. Dimensions can be held to an accuracy of about 0.008 to 0.01 in. per in., and casting surfaces are excellent, often requiring neither machining nor grinding.

Plaster mixes for metal casting ordinarily contain 20 to 30 per cent talc to prevent mold cracking, and may contain compounds such as terra alba or magnesium oxide to hasten setting time, or other compounds to retard setting time. Materials such as lime or cement can be added to control expansion of the plaster during baking. Chemical changes in the mixing and baking, however, involve only the water of crystallization of the gypsum. As received, calcined gypsum is $CaSO_4 \cdot \frac{1}{2}H_2O$; during the initial *set,* it reacts with the water of the slurry to form $CaSO_4 \cdot 2H_2O$. When dried at temperatures below about 320°F (160°C), it reverts to $CaSO_4 \cdot \frac{1}{2}H_2O$, and, at temperatures above 320°F (160°C), the last combined water is driven off, leaving anhydrous calcium sulfate ($CaSO_4$). Because molds of metal casting plaster possess very low permeability, care must be taken to remove *all* combined water, and avoid absorption of moisture after baking.

An alternative plaster casting technique, known as the *Antioch process* has been applied successfully for making special engineering parts of complex shape requiring fine detail and thin sections. The major advantage of the Antioch process is that it develops a high degree of permeability in the plaster mold, making it easier to obtain fine detail and allowing any moisture or other gases present to escape. To make molds by this process, water is added to a dry mixture of gypsum, sand, asbestos, talc, and sodium silicate to make a slurry which is piped by hose into metal coreboxes or cope and drag flasks fitted to special matchplates. The mixture develops an initial set, patterns are drawn, and the mold is assembled in the green condition. After standing 6 hours at room temperature, the molds are placed in a steam autoclave at about 2 atmospheres' pressure. They are then cured in air for 12 hours and finally oven-dried at temperatures up to about 450°F (230°C). Autoclaving develops a special permeable structure in the mold but greatly reduces dry strength. The oven-drying cycle drives off the free and combined water. Because of the high mold permeability, it is not always necessary to drive off all the combined water.

Another type of plaster mix, known as *permeable metal casting plaster* has been designed to produce mold permeabilities comparable to those obtained in the Antioch process, but without the autoclaving step. In this process, a *foaming agent* is added to ordinary metal casting plaster, and air is beaten into the mix with a rapidly rotating rubber disk. Setting and baking steps are similar to those of the metal casting plaster process, except that baking may be at lower temperatures and for shorter times since the high mold permeability makes it unnecessary to drive off all combined water. A properly made mold of permeable metal casting plaster contains as much as 50 per cent air holes finely distributed throughout the mold (except for a thin layer at the pattern surface). Permeability of the mold can be adjusted by varying the amount of air beaten into the mix.

The advantages of any plaster casting process are that nonferrous castings having intricate and thin sections can be made with good dimensional accuracy and excellent surface finish. The Antioch and permeable metal casting plaster processes possess the additional advantages of mold permeability and the ability to incorporate chills in the mold. Chills cannot be used as readily in molds of ordinary metal casting plaster since they tend to expand and crack the brittle mold during baking.

Plaster casting is suitable only for making non-ferrous castings; the sulfur of the gypsum reacts chemically with ferrous metals at high temperature to give very bad casting surfaces.

3.3 Investment casting

Investment casting (Fig. 3.2) is known variously by the terms *lost wax, lost pattern, hot investment,* and *precision casting.* The method antedates Michelangelo and Cellini, who used it for statuary and art work. For many years the process was neglected but it has now come into favor with dentists, jewelers, and orthopedic surgeons. It was adapted, particularly during World War II, to the manufacture of engineering castings of many types from metals

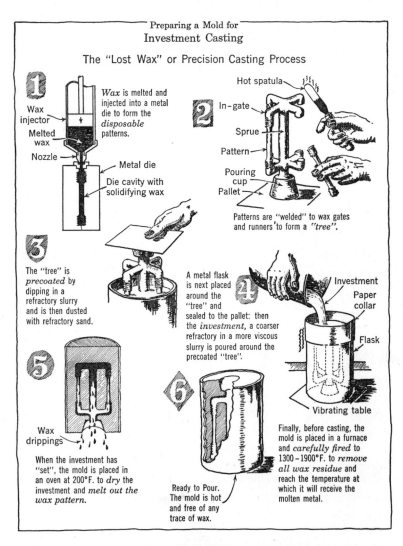

Preparing a Mold for
Investment Casting

The "Lost Wax" or Precision Casting Process

1 *Wax* is melted and injected into a metal die to form the *disposable* patterns.

Wax injector
Melted wax
Nozzle
Metal die
Die cavity with solidifying wax

2 Hot spatula
In-gate
Sprue
Pattern
Pouring cup
Pallet

Patterns are "welded" to wax gates and runners to form a *"tree"*.

3 The "tree" is *precoated* by dipping in a refractory slurry and is then dusted with refractory sand.

4 A metal flask is next placed around the "tree" and sealed to the pallet: then the *investment*, a coarser refractory in a more viscous slurry is poured around the precoated "tree".

Investment
Paper collar
Flask
Vibrating table

5 Wax drippings

When the investment has "set", the mold is placed in an oven at 200°F. to *dry* the investment and *melt out the wax pattern.*

6 Ready to Pour. The mold is hot and free of any trace of wax.

Finally, before casting, the mold is placed in a furnace and *carefully fired* to 1300–1900°F. to *remove all wax residue* and reach the temperature at which it will receive the molten metal.

FIG. 3.2. Preparing a mold for investment casting.

which were not amenable to fabrication by other than casting methods.

In this process an *expendable pattern* of *wax, plastic, tin,* or *frozen mercury* is used. The pattern (patterns, if several castings are to be made at once) is prepared by attaching suitable gates and risers, and the assembly, or *tree,* is placed inside a container, usually a stainless-steel cylinder open at each end. A slurry of suitable binder plus alumina, silica, gypsum, zirconium silicate, or mixtures of these and other refractories is then poured over the pattern and the whole vibrated to remove air bubbles. After the refractory has taken an initial set, the container is placed in an oven at low heat; the refractory becomes harder, and as the temperature of the furnace is raised steadily the pattern either melts and flows from the mold if it is wax, or volatilizes if made of a plastic such as polystyrene. The mold now contains a cavity in the identical form of the original pattern; the temperature is raised to 1200 to 1800°F (650 to 1000°C), and molten metal is poured into the hot mold. Any alloy that can be melted is amenable to investment casting by adapting refractory and mold temperatures to the requirements of the metal being poured.

The basic method has many modifications. For example, patterns are sometimes dipped in a slurry and dusted with fine sand before *investing;* the slurry used for this process imparts a fine-textured surface at the mold-metal interface, and the sand serves to *key* the precoat to the regular investment. In a new technique, refractory costs are minimized by forming only a thin shell of the refractory around the pattern. The shell is formed by dipping the pattern several times into a rapidly setting refractory slurry (until a shell of desired thickness has been built up around the pattern). The thin refractory mold is then given a dry stucco finish; subsequent baking and firing steps are similar to those of ordinary investment casting.

Investment materials can also be used to make cope and drag molds (*cope and drag investment casting, ceramic casting*). In this case reusable metal patterns are employed, and the procedure is similar to that of plaster casting except, of course, that baking temperatures and times are different. Cope and drag investment casting is used in preference to lost-pattern investment casting (1) when castings are too large to be produced economically by the latter process, and (2) when the parting line resulting from the cope-drag combination is not a serious disadvantage. The major disadvantage of a parting line is that it is not as easy to control dimensions across a parting line as in the absence of such a line; also, the trace of the parting line is usually visible on the casting, and this may be undesirable from the standpoint of casting appearance or fatigue life. Cope and drag investment casting is used in preference to ordinary sand casting techniques (1) to produce castings with improved surface finish and dimensional control, and (2) to minimize certain foundry defects (such as sand inclusions and hot tearing) in intricate high-quality castings.

In *lost-pattern investment* casting, wax patterns are usually preferred to other types because of relative cheapness and because *treeing* can be done easily with a warm spatula. Dimensional accuracy is not quite as good as for tin or plastic, and warm weather sometimes causes sagging. To increase accuracy in using wax for patterns, an air-conditioned room should be provided. In order to obtain optimum dimensional control, it is necessary to make certain allowances in pattern making; the wax may change dimensions slightly, depending upon its composition, since it is cast in a die from liquid form; the mold changes dimension in being heated to high temperatures; and the cast metal will contract upon solidification. All these dimensional changes may be determined and compensation can be made in the die from which patterns are made.

Molds may be filled by gravity or by applying pressure to the molten metal as it fills the mold; this may be done by direct air pressure or centrifugal force. Since the molding material is relatively very dense, special air vents are usually provided to allow mold vapors and other gases to escape. Molds are not always poured hot; at some sacrifice to surface finish and dimensional accuracy, castings may be poured in cold or slightly heated molds, either by gravity or by pressure.

The chief characteristics of lost-pattern investment casting are (1) production of castings with extreme dimensional accuracy and excellent detail and surface finish, (2) accommodation of unmachinable and unworkable alloys for such uses as turbine blades, jet nozzles, gun and machine parts, and parts for household and other appliances, (3) no disfiguring parting line as found on castings made by cope and drag methods, and (4) high cost.

3.4   Mold materials for investment casting

Fine silica sand and silica flour are frequently used as the base ingredients of investment casting mixes. Other refractories, however, are often used because of their higher fusion point or better thermal stability; these include alumina, magnesia, and zircon (see Table 2.3). Gypsum may be used for the lower-melting non-ferrous metals. Binders for use in investment casting mixes include ethyl silicate, magnesium phosphate, furfuryl alcohol, and gypsum. Each of these binders develops an initial *set* after pouring and is later baked at higher temperatures to drive off moisture or other volatile residue.

A promising new development in investment casting is the "glass cast process." In this process finely ground *Vycor glass* (a silicate) serves as the mold refractory. Vycor has an extremely low thermal coefficient of expansion and does not undergo phase transformations on heating, as does quartz. A low-temperature binder is used to obtain "green" strength. After high-temperature baking, the "green" binder burns out completely, and the Vycor bonds itself by electrostatic attraction of the fine particles.

3.5   Die and permanent mold casting

Die casting and *pressure die casting* are terms describing identical processes in America and England respectively; the terms are used interchangeably to describe both the process and the product. Die casting is the production of castings *under pressure* in *permanent metal molds;* it is distinguished from the English *gravity die casting* and the American *permanent mold casting* in which permanent metal molds are also used, but the metal fills the mold under *gravity* or *centrifugal* pressure only. In both the die and the permanent mold casting processes the molds are reused many times.

Permanent mold castings are much thicker in cross section than die castings, as gravity alone does not provide enough pressure to force molten metal into metal molds in sections much thinner than $\frac{1}{4}$ or $\frac{1}{2}$ in. Permanent molds are made in two halves; they may be designed with a vertical parting line, as in Fig. 3.3, or with a horizontal line as in ordinary sand molding. The mold material is usually a good grade of cast iron, although die steel, graphite, copper, and aluminum have also been used. Figure 3.3 illustrates a typical permanent mold for casting aluminum bearings; the casting with gate, riser, and core is shown in the left half of the mold. Cores for permanent molds can be sand, plaster, collapsible metal cores, or simply heavily tapered metal cores which are removed while the casting is still hot.

Permanent molds may be used singly or mounted on turntables for production work. The sequence of operations includes (1) cleaning the molds by brushing or blasting with warm air, and maintaining them at proper casting temperature by a gas or oil flame; the correct operating temperature can be determined only by experience and varies with the casting; (2) painting or spraying the mold surface with a thin refractory *wash,* or blacking it by depositing carbon from a reducing oil or gas flame; (3) inserting cores, if used, and closing the mold by hand or by automatic action; (4) pouring the metal from a hand ladle or a bull ladle suspended on a movable track; (5) allowing sufficient time for the casting to solidify; and (6) ejecting the casting from the mold automatically or by hand. Permanent mold castings have been made commercially of tin, zinc, lead, aluminum, magnesium, copper, and cast iron, and from their alloys. Higher production rates are obtained in permanent mold casting than sand casting, but the process is much slower than die casting.

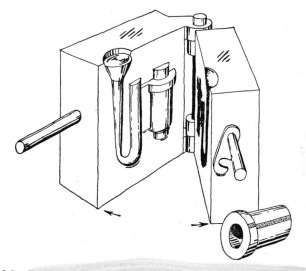

Fig. 3.3.   Permanent mold casting.   Solidified casting with gate and metal core shown in left half of mold with completed casting in front.

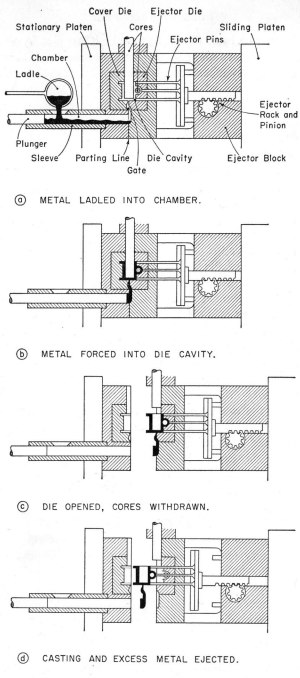

(a) METAL LADLED INTO CHAMBER.

(b) METAL FORCED INTO DIE CAVITY.

(c) DIE OPENED, CORES WITHDRAWN.

(d) CASTING AND EXCESS METAL EJECTED.

FIG. 3.4.   Die casting, cold-chamber machine.

59

There are two basic die casting methods, typified by the names given the machines: *cold chamber* and *gooseneck*. Figure 3.4 explains the cold-chamber process schematically, as it would apply to the manufacture of a cup-shaped piece. The molten metal, which may be an alloy of zinc, aluminum, copper, magnesium, lead, or tin, is ladled from the melting pot into a pouring slot in the main pressure chamber (Fig. 3.4*a*). A close-fitting plunger is rammed forward by hand or pneumatic action; this confines the metal against the die opening and actually squirts it into the die at pressures of 20,000 psi and higher (Fig. 3.4*b*). Because most die castings are of thin section (usually $\frac{1}{8}$ in. or less), solidification is extremely rapid. After the casting is solid, the die halves are opened and cores withdrawn (Fig. 3.4*c*). Finally, ejector pins automatically force the casting out of the cavity (Fig. 3.4*d*).

The principle of the gooseneck machine is shown in Fig. 3.5. Pressures usually do not exceed 600 psi and may be applied directly by metal plunger or by air. The gooseneck may be filled by hand, as in the cold-chamber method, or it may be actuated by a cam to alternately dip into the metal and rotate into position against the die opening before pressure is applied for the shot. Air is preferred to direct metal plunger for higher-melting alloys like those of aluminum and copper; with these metals the plunger erodes relatively rapidly, requiring replacement, and iron may contaminate the casting and impair its strength, ductility, or corrosion resistance. For aluminum and copper alloys, the cold chamber is the more widely used die casting method.

Metal for die casting is usually melted in pots directly from ingots; sprues, runners, and *flash* from castings made previously are also added. In some shops metal is melted in large furnaces and transferred to smaller pots for temperature regulation and special alloying preparatory to use. Dies are elaborately water-cooled to maintain them at the lowest temperature commensurate with satisfactory mold filling; this prolongs die life and provides the fastest allowable cooling rate for castings in order to develop optimum properties and minimum lost time. Sometimes die castings are quenched in water directly from the die to improve properties. Cores in dies may be loose (knockouts) or fixed; they may be actuated automatically as shown in Fig. 3.4, or *set* by hand for each operation if deep recesses are required. Castings may be drawn by hand from the die as soon as it is opened, or may be removed instantaneously by automatic *ejector* pins. Nearly all castings have fins called *flash* where metal has flowed past cores and between die halves at the parting line.   Flash is removed by *grinding* or *blanking,* i.e., press shearing.   Dies for tin and lead alloy casting are made of carbon steels without heat treatment; for zinc, aluminum, and magnesium, dies are of heat-treated low-alloy steel; and for copper-base castings heat-treated special alloy steels are used.

The die casting process is essentially a high first-cost, high-production-rate process.   Die casting machines cost upwards of $25,000.   A set of dies for a casting run may cost from $2500 to $30,000.   Production rates, however, are usually between 100 and 400 shots an hour, and up to forty castings may be made in a single shot (40-impression die).   It is generally considered that somewhere in the range of 50,000 to 100,000 castings must be produced to make the die casting process economical.   The process is usually used only for small castings of relatively thin section, although castings up to 40 lb. (automobile engine blocks) have been experimentally produced in this manner.   Die castings may be machined or coined (pressed) for finishing or for slight changes of shape, or may be polished, plated, dipped, sprayed, treated chemically, coated with plastic, enameled, or lacquered directly from the casting operation.

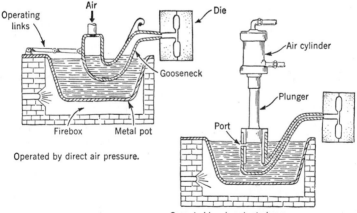

FIG. 3.5.   Die casting, gooseneck machine.

One disadvantage of the die casting process has been that air or other gases tend to be entrapped in the casting in fine bubbles. These gas holes can result in low ductility and impact strength, and can expand during subsequent heat treatment to cause *blistering*. In a new technique, the so-called *vacuum die casting process,* the die cavity is evacuated before each shot. Evacuation is accomplished either by (1) sealing the die halves as they close and applying a vacuum directly to the casting cavity, or (2) enclosing the entire die assembly in a vacuum-tight container which closes and seals when the die closes. *Vacuum die casting,* properly controlled, can result in sounder castings than are ordinarily obtained in die casting; however, more work remains to be done before this technique will be fully dependable.

In another new die casting technique, small parts are cast integrally into movable assemblies. For example, complete watch chains, small scissors, binder rings that open and close, etc., can be cast complete in one operation. In the case of scissors, one blade is first poured; the cast metal solidifies rapidly and forms part of the die for the second blade. Metal for the second blade enters the die cavity a split second later and flows around the first material and into a pivot hole in the solidified first blade. The result is a pair of scissors, pinned and movable without further operations.

3.6 Centrifugal casting

Centrifugal force has been used in a variety of ways in the foundry industry as a mold-filling device, for making tubular objects, and for purifying metals. Since its invention at the beginning of the nineteenth century, centrifugal casting has enjoyed the attention of engineers and promoters for brief periods of a few years at a time, but only a limited number of applications developed have survived commercial exploitation. All the advantages and disadvantages of the various processes have never been clearly defined in an engineering sense, but certain observations can be made. (1) Centrifugal casting has proved definitely economical for making tubular objects. No core is needed to form the bore as in static casting. (2) Centrifugal pressures can be applied to advantage in forcing molten metal quickly into molds to prevent premature freezing and to aid slightly in controlling temperature gradients.   (3) The conditions of solidification in centrifugally cast tubes, particularly those cast in chill molds, develop optimum properties in the casting, but other types of castings made centrifugally are not superior to well-made static castings.   (4) The casting must be ideally suited to the process or it can probably be made to better advantage by static methods.

Centrifugal casting is divided into three categories (Fig. 3.6): (1) *True centrifugal casting,* in which the casting is spun about its own axis; no risers are required and no central core is needed since centrifugal force forms the inner diameter of castings such as pipe naturally; (2) *semicentrifugal casting,* in which the object, such as a wheel with spokes, is spun about its own axis, but risers and cores are needed; and (3) *centrifuged casting,* wherein the mold impressions are grouped around a central downgate, as in static casting, and centrifugal force is used mainly as a mold-filling device.

Several methods for true centrifugal casting are important commercially.   In the *Watertown method,* by which large gun tubes are made, a heavy cast-iron rotor is used as a chill mold.   The inner diameter of the rotor is machined to produce a gun tube of definite outside diameter.   The mold is coated with refractory wash and spun at 1300 rpm.   When the mold is at speed, a weighed amount of metal is poured into a pouring basin equipped with a short spout at the large end of the bore.   The rate of entry of metal into the mold is accurately adjusted to spinning speed by regulating the spout diameter so that molten metal will be distributed evenly and without tearing the solidified, but hot, shell. If the rate of entry is too slow, surface laps result from premature freezing; if too fast, the casting will crack because the force of molten metal on the solidified thin skin next to the chill mold will be too great.   Another feature of this process is that pressure (about 75 psi) is applied against the small end of the tube shortly after solidification to prevent mold restraint and consequent tearing as the tube contracts lengthwise in the mold.

In the *DeLavaud method,* used for making soil pipe and similar tubular objects, a relatively thin-walled metal mold is provided with a water jacket for cooling.   The mold and spinning equipment are mounted on a track inclined at a slight angle to the horizontal.   A pouring trough, sometimes as long as 16 ft, is extended into the spinning mold. Molten metal is deposited in a helix from the farthest end of the trough as the spinning mold is retracted. Water pipes as large as 12 in. in diameter and 20 ft long are made by this method.

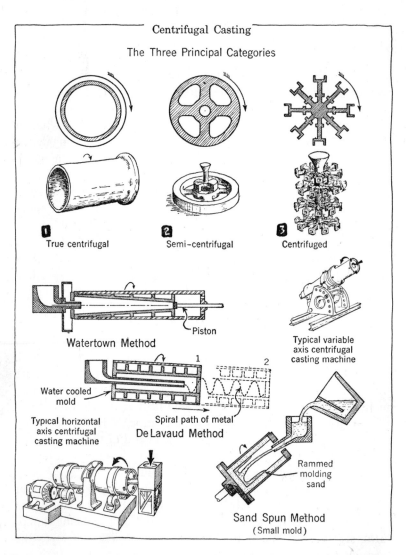

Fig. 3.6. Centrifugal casting.

The *sand-spun process* is essentially the same as the Watertown method except that a sand lining is rammed inside a steel shell. The chilling rate of the metal is relatively much lower than for other methods.

One modification of the Watertown system employs a much thinner metal mold which is equipped with spray cooling. Mold life is considerably improved and the cost reduced as compared to the use of extremely large rotors.

In World War II, the Germans made some notable contributions to centrifugal casting. Instead of coating the mold with mold wash, a layer of fine, dry, unbonded sand was deposited from a long trough over the mold surface; molds were made of metal and were spray-cooled. A streamlined pouring device was developed to deposit molten metal gently at one end of the tube; the layer of loose sand was not disturbed, and sand grains were not washed into the metal, even though the density of sand is only about one-third that of steel. The thermal barrier provided by the sand layer reduces heat shock, lessens checking of the mold surface, and prevents surface laps on castings caused by premature freezing. The effects of the thermal barrier disappear relatively quickly after the mold is partially filled, allowing the metal to cool rapidly.

3.7 Slush casting

For some castings, such as statuary work, only the external features of the casting are important (for their esthetic value). Since such castings are not destined for engineering use, uniformity of wall thickness is not an important consideration. A core is not necessary—instead the mold is filled with molten metal and held stationary until a thin skin of solid metal freezes against the mold walls; the mold is then inverted and the unfrozen metal "bled" from the casting. The mold must not be inverted too sharply or air will be excluded as molten metal chokes the cavity; a partial vacuum will be formed, and the still weak solid skin of the casting will collapse at its weakest point. Also, the casting must be made of a relatively pure metal, as most alloys do not form a strong solid skin; this is discussed in detail in Chapter 5.

Slush casting of metals corresponds to slip casting techniques used by ceramists to make ceramic bodies. It is used only for a limited amount of art and decorative work. Perhaps the most

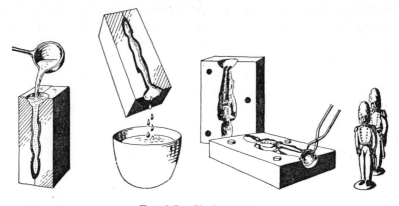

FIG. 3.7.   Slush casting.

familiar example of a casting of this type is the "tin" soldier. Figure 3.7 illustrates the process of slush casting a tin soldier in a permanent mold.

3.8   Shot casting

Metal shot for use in shot-blasting equipment is made by dropping molten metal from considerable heights into a pool of water. The molten metal is flowed over a control weir at a fixed rate so the thin stream separates into droplets (because of surface tension forces); the droplets become spherical in shape as they fall freely through air and freeze in this form upon striking the water.   The size of shot so produced is regulated by type of metal, pouring temperature, rate of exit of the metal stream, distance through which the droplets fall, and any air jet or mechanical device used to disperse the stream as it exits from the control weir or nozzle. In Civil War days shot towers were constructed like giant chimneys, from the top of which molten iron was poured into tubs of water. Grape shot (irregularly shaped, roundish balls of metal) were made by this technique; manufacture of large cannon balls was attempted by this method but could not be produced spherical enough to use.

3.9   Summary

To compete commercially, a new casting process must either make a better casting, make a given casting cheaper, or make a part which cannot be made by another process: e.g., by investment casting one can make turbine blades of unmachinable or unworkable metals; die casting mass-produces parts relatively cheaply; and centrifugally cast gun tubes are better than forgings.

1. *Plaster casting* is described as a process employing a special mixture of plaster of paris (gypsum, $CaSO_4 \cdot nH_2O$, plus asbestos and talc). The molding material is available as *metal casting plaster;* when mixed in proportions of 100 parts to 160 parts water, poured over a metal pattern, set, and dried, an excellent mold is produced. In the *Antioch process* it is mixed with sand or additional asbestos, and *autoclaved.* Any process involving gypsum is *suitable for making only non-ferrous castings;* the sulfur of $CaSO_4$ reacts readily with *ferrous* materials producing castings with rough surfaces.

Advantages of the plaster casting process are that non-ferrous castings of intricate shape can be made with good dimensional accuracy and excellent surface finish.

2. *Investment casting* provides the best control over dimensions and surface finish of any process known. An *expendable pattern* of wax, plastic, tin, or frozen mercury is used, and castings bear no disfiguring parting lines as do castings made by regular cope and drag processes. An *investment* is made of the pattern by coating it with a slurry of refractory; the investing material is allowed to *set* and is then heated until the pattern melts or vaporizes. The mold is then heated to red heat or above and metal is poured into the mold under pressure. Processing costs are relatively very high, but since no machining or other surface finishing is required many articles, particularly those made of very hard metals, can be made economically by this process.

Chief characteristics of investment casting are (1) excellent dimensional accuracy, and good detail and surface finish of castings, (2) wide latitude of composition in parts made by the process, (3) lack of parting line and other blemishes on castings, and (4) relatively high cost.

Newer investment casting processes permit a saving in the amount of refractory required by employing a thin "shell"-type refractory mold. Another process, *cope and drag investment molding,* permits achieving some of the advantages of investment casting without the use of expendable patterns; one such process is the "Shaw process."

3. In the past, die casting has meant mass-producing simple, thin-walled parts of zinc-, tin-, or lead-base alloys under pressure in permanent metal molds; today the range of alloys includes aluminum-, copper-, and magnesium-base alloys, and complex parts such as sewing machine and sander frames, whole automobile doors, and typewriter frames are made at the rate of one every 2 to 5 minutes. *Die casting* is distinguished from *permanent mold* casting in that in die casting the metal is forced into a permanent metal mold under air or piston pressure, whereas only the force of *gravity* is used for filling the metal molds in *permanent mold* casting; a less confusing distinction is to call the former *pressure* die casting and the latter *gravity* die casting.

The maintenance of dies is a limiting feature to the possibility of die-casting iron-base and other high-melting alloys; the combination of repetitive heat shock and high pressures cause rapid breakdown of die surfaces even when used for aluminum- and copper-base alloys. In permanent mold (or gravity die) casting production is relatively much slower than in die (or pressure die) casting, and dies do not suffer drastically from heat shock and pressure; therefore castings of iron-base alloys (steels and irons) are made satisfactorily by this process.

4. *Centrifugal casting* is a popular and intriguing process to engineers but only a relatively small proportion of the total tonnage of castings are made this way. It is a popular misconception that all castings made centrifugally are *inherently* better than properly made static castings. Gun tubes, and other cylindrical parts, which can be spun about their own axis with no core are often superior to similar parts made by static methods because directional solidification is ideal and effective. This is usually a process employing high speed (1300–2000 rpm) and is called *true centrifugal casting* to distinguish it from *semicentrifugal* and *centrifuged* casting; in semicentrifugal casting, typified by spinning a wheel with spokes about the axis of the hub, typical risers are used and spinning speed is relatively low (200–500 rpm). In *centrifuged* casting several relatively small objects are spun about the axis of a downgate which leads outward into each individual casting. In the latter two processes the force of spinning is utilized mainly as a *mold-filling* device.

Modifications of the true centrifugal process are (1) Watertown, (2) sand-spun, (3) DeLavaud, and (4) German processes. Attempts to *purify* molten metal by centrifuging, in cream-separator fashion, have never proved successful. For castings ideally suited to one or another of the centrifugal casting processes, economics and quality justify the relatively high overhead and maintenance costs of the method; economic factors and degree of *suitability* seem to be the main deterrents to wider industrial application of this technique.

5. *Slush casting* and *shot casting* are two processes with a rather limited range of usefulness. Slush casting is employed when only the exterior appearance of a casting is of importance; the casting is allowed to partially solidify, and the central core of still liquid metal is then dumped out. Shot casting (dropping fine droplets of metal into a quenching bath) is used to produce metal shot of a wide variety of sizes for such purposes as shot blasting.

6. *Ingot making,* a most important casting process, has not been considered in this chapter. The choice of an ingot casting process, and the engineering of that process are determined in large part by the *solidification characteristics* of the metal to be cast. Discussion of ingot making has been deferred until after solidification and risering are considered; the ingot as a casting is discussed in Chapter 8.

3.10 Problems

1. List the various casting processes and one unique property which distinguishes each process from all other processes.

2. What precautions must be taken in the plaster casting process?

3. Write the chemical reactions for hydration and dehydration of gypsum. Why are such high drying temperatures required in plaster molding?

4. List advantages of the Antioch process as compared to other casting processes. Also list the disadvantages.

5. Describe the precision casting process briefly.

6. List advantages and disadvantages of the precision casting process.

7. What factors must be considered in using the die-casting process? the permanent mold process?

8. What are the essential differences between die casting and permanent mold casting?

9. What types of alloys are used in die casting and permanent mold casting? What factor limits the use of these processes for any and all alloys?

10. What principles are involved in centrifugal casting? What types of objects are best made by this process? Are centrifugal castings inherently superior to static castings?

11. What are the advantages of the true centrifugal casting process?

12. How does true centrifugal casting differ from semicentrifugal and centrifuged casting?

13. Sketch typical cold-chamber and gooseneck die-casting methods.

14. Indicate the casting process most applicable to making the following: (1) class rings, (2) soil pipe, (3) spindle for a record changer, (4) sewing-machine housing, (5) large piston rings, (6) aluminum ash trays, (7) bronze ash trays.

15. Do you think steel castings can be made in permanent molds?

16. Obtain one or two molds for making so-called "lead soldiers" or other toys (or make a simple metal mold for casting lead "sinkers" for fishing). Then see if you can improve on the surface appearance of the castings you make by (a) coating the mold with carbon black from a lighted candle, (b) coating with a mixture of carbon black in some suitable liquid (and drying, of course), (c) polishing the mold to a smoother surface condition and more accurate "fit."

17. Locate some large piece of statuary near your school and determine (a) the name of the sculptor and the date of casting, (b) the name and address of the foundryman who did the casting, and (c) all essential details of the method of sculpturing and the casting process. This may not be as easy as might be thought unless the statuary is relatively new.

18. Obtain a piece of "foamed" polystyrene and shape it into a simple block $6 \times 4 \times 2$ in. Ram molding sand around this "pattern," and cut a downgate over one end and a ¼-in.-diameter vent hole over the other. Do not bother to remove the pattern from the mold but, instead, simply melt some aluminum and pour into the mold. The resulting casting will be better than you might expect. What industrial or art uses can you think of for such a process?

19. Write a 1000–2500-word term paper on why the classical "sand-cast, cast-iron engine block" might be replaced by an aluminum die casting. It is suggested you write to General Motors Corporation and Reynolds Metal Company for reference material.

20. Make a table of the composition and cost (price per ton F.O.B. factory or warehouse) of typical mold materials (include the following):

(a) Silica sand (natural).

(b) Silica sand (washed and graded).

(c) Zircon sand.

(d) Olivine.

(e) Chamotte.

(*f*) Aluminum.

(*g*) Cast iron.

(*h*) Aluminum oxide.

(*i*) Magnesium oxide.

(*j*) Gypsum.

3.11   Reference reading

1. *Die Casting for Engineers,* New Jersey Zinc Co., New York, 1946.

2. "Symposium for Centrifugal Casting," *A.F.A. Publication* **44–37,** Chicago, 1944.

3. Doehler, H., *Die Casting,* McGraw-Hill, New York, 1951.

4. Wood, R. L., *Investment Casting for Engineers,* Reinhold, New York, 1952.

5. *Product Information, Data, Applications, and Use,* U. S. Gypsum Co., Chicago.

CHAPTER 4

The Nature of Cast Metals

PART I. BASIC METALLURGY OF CAST METALS

4.1 Introduction

At this point we turn our attention from the strictly molding and mold-making aspects of foundry to the *cast metal* itself. Here we find an area which historically has been shrouded in deep mystery. It was not so long ago that metal was considered to shrink because it was too "male," or had certain other characteristics because it was too "female." Strange herbs and animal extracts were essential to the production of sound castings; the ability to produce castings at all was a "God-given art," bestowed on a chosen few.

Advances of metallurgy in the last 50 years and less, however, have given us a rational understanding of many of the variables affecting cast metals and the casting of metals. It is the purpose of this and the following chapter to present the metallurgy of founding as simply and concisely as possible. Subsequent chapters deal with the practical applications of this metallurgy in risering, melting, heat treatment, and other foundry processes.

4.2 The crystalline nature of metals

As a molten metal is cooled to its freezing temperature, solidification begins in the region or regions of lowest temperature. Submicroscopic metal crystallites, called *nuclei* first form and then grow, generally in pine tree or *dendritic* fashion, as sketched in Fig. 4.1. The growing arms of the dendrites eventually impinge, one upon another, and further growth takes place by formation of small arms upon the original branches of individual dendrites until the mass is completely solid. The solid dendrites are alternatively called *grains* or *crystals*.

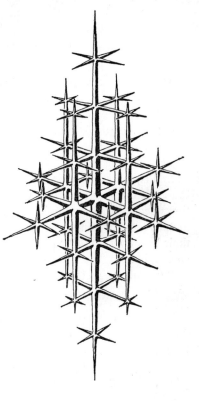

FIG. 4.1. Growth of a metal dendrite.

As the metal atoms attach themselves to the solid to form the growing arms, they do so not randomly but with an amazing degree of order, or *crystallinity*. The atoms line themselves throughout the entire grain in evenly spaced columns and rows so that the final structure may be pictured as many *unit cells* stacked one upon another. Some types of unit cells are sketched in Fig. 4.2. Austenitic iron, copper, and aluminum possess the face-centered cubic structure; ferritic iron and chromium are body-centered cubic, and magnesium and zinc are hexagonal close-packed. These differences give unique properties to the metals of each class, but it is not within the scope of this book to discuss them; the characteristics arising from the differences in basic structure are very important in metal working but are not particularly significant to foundry processing.

4.3   Characteristics of pure metals

Pure metals possess certain characteristic properties which make
them desirable for specialized applications. They have very high
electrical conductivity; hence, commercially pure copper or aluminum are used for electric wiring, bus bars, or other carriers of
electric current. Pure metals also have excellent thermal conductivity, and so commercially pure copper is used for the tuyères
in water-cooled blast furnaces. Other characteristic properties of
nearly pure metals (as compared with their alloys) are, typically,
higher melting point, higher ductility, lower tensile and yield
strengths, and (sometimes) better corrosion resistance. In the
foundry, the generally higher melting point of pure metals sometimes leads to difficulty in pouring castings, more severe mold-
metal reactions, and greater tendency toward cracking. Further,
their *mode of solidification* often renders them difficult to make
sound, i.e., to "riser" properly.

Pure metals melt and freeze at a single temperature, the *melt-*

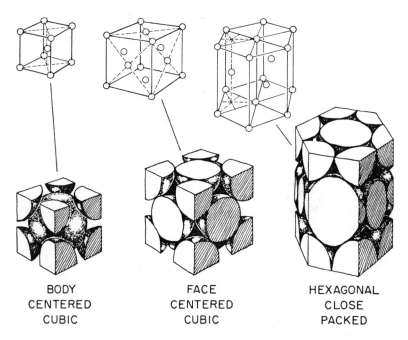

BODY	FACE	HEXAGONAL
CENTERED	CENTERED	CLOSE
CUBIC	CUBIC	PACKED

FIG. 4.2. Some common stacking arrangements of metal atoms (unit cells).

ing point (or *freezing point*); above this temperature they are completely liquid, and below completely solid (under equilibrium conditions). If a thermocouple is placed in a crucible of pure molten metal and the metal allowed to solidify slowly, a cooling curve such as that of Fig. 4.3*a* is obtained. It will be observed that the metal cools quite quickly to its freezing point. At the freezing point the temperature remains constant briefly; i.e., a *hold* occurs while the metal loses its heat of fusion. Only after the metal is completely solid can further cooling occur.

When pure metals are cooled rapidly, formation of solid crystals at the freezing point is sometimes inhibited. In such cases a cooling curve such as that of Fig. 4.3*b* is obtained. At a temperature (ΔT) below the freezing point, nucleation of solid suddenly occurs, and the heat of fusion liberated may increase the temperature to that of the equilibrium freezing point. Such cooling of a liquid metal below the normal solidification temperature is called *under-cooling* or *supercooling*. Supercooling also can occur in commercial alloys, and has a direct bearing on the metallurgical structures of these alloys; for example, the size, shape, and distribution of graphite in gray cast iron can be directly related to "supercooling" effects.

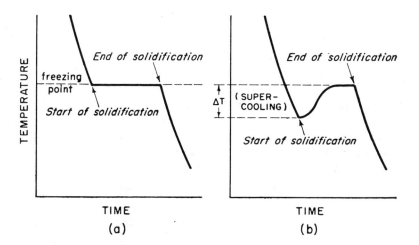

FIG. 4.3. Cooling curves of a solidifying pure metal. (*a*) Equilibrium cooling, (*b*) Cooling curve showing supercooling.

4.4   Characteristics of alloys

While pure metals are desirable for certain specialized applications, *alloying elements* are usually added to cast metals to modify or enhance their properties. Alloyed metals, or *alloys,* are generally stronger than pure metals; the tensile strength of pure iron, for instance, may be improved by a factor of six or more by proper alloying and heat treatment. Alloying elements are also added to improve such diverse properties as strength at high temperature, machinability, corrosion resistance, color, and others.

Alloying elements may be added to metals solely to improve the foundry characteristics of the metal; they usally lower the melting point of a metal, and alter its mode of solidification. The effect of such additions on the mode of solidification is discussed in Chapter 5. It is sufficient here to state that alloying elements exert a profound effect on the ease with which a given metal may be cast. Compare, for instance, the castability of low-carbon steel and gray cast iron. The amount of carbon in steel is so small that steel solidifies very much like a pure metal. It has a high melting point, flows with relative difficulty, tends to erode and react with the mold, and is prone to hot tearing. Further, the mode of solidification of steel makes it difficult to "feed" properly during solidification. On the other hand, the addition of 3 or 4 per cent carbon and silicon, results in a material (cast iron) which freezes altogether differently and several hundred degrees lower. Mold-metal reactions, hot tearing, and risering difficulties are all very much reduced. Gray iron is considered an "easy" alloy to cast; steel a "difficult" alloy.

Alloying elements usually lower the melting point of a pure metal. In addition, such elements increase the range over which melting occurs. Pure metals melt and freeze at a single temperature. Alloys, in most cases, do not. For a given alloy, there is a particular temperature (the *liquidus* temperature), above which it is all liquid. There is another, lower temperature (the *solidus* temperature), below which it is all solid. Between these two temperatures is a region where liquid and solid coexist. Such an alloy, in this temperature range, has some consistency because of the solid material existing, but little or no strength, because of the liquid interspersed with the solid. It has the consistency of "mush," and for this reason the portion of a solidifying casting

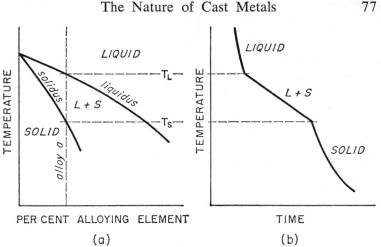

FIG. 4.4. Effect of alloying on the solidification temperature of a metal. (*a*) Plot of solidification temperature versus per cent alloying element, (*b*) Ideal cooling curve (equilibrium cooling) of alloy.

whose temperature lies between the liquidus and solidus is often termed the "mushy zone"; this is described in detail in Chapter 5. A sketch of the liquidus and solidus temperatures of typical alloy series is shown in Fig. 4.4*a*.

When a molten alloy such as alloy *a* of Fig. 4.4*a* is slowly cooled, freezing begins at T_L but is not complete until a lower temperature, T_S, is reached. The cooling curve of an alloy of this type would be very different from that of a pure metal. If a crucible of the alloy were cooled slowly and temperature measured as a function of time, the rate of cooling would be rather rapid until the liquidus temperature, T_L, was reached. At this temperature, as solidification began and continued over a range of temperatures, the cooling rate would be slowed down, but not arrested. Below the solidus the metal would be completely frozen, and more rapid cooling would again take place (Fig. 4.4*b*).

4.5 Solid solution alloys

One of the ways in which two metals may combine to form an alloy is by *solid solution*. Several types of solid solutions are possible. In steel, the carbon atoms wedge themselves in the interstices between the iron atoms; they "dissolve" *interstitially* in the iron crystal structure. A more common type of solid solution, however, is represented by the copper-nickel alloys. Both copper and nickel are face-centered cubic metals. If an alloy is composed of 50 per cent copper atoms, and 50 per cent nickel atoms, the alloy will have a single crystal structure which is also face-centered cubic. One-half the atom sites in the crystal structure will be filled with copper atoms, and one-half with nickel atoms. The metals, in effect "dissolve" in each other's crystal lattice to form a *solid solution*.

The liquidus and solidus temperatures of copper-nickel alloys can be plotted versus per cent copper as shown in Fig. 4.5. Such a plot constitutes a *phase diagram* or *equilibrium diagram* for the system. Note that copper and nickel are mutually soluble in all proportions. All compositions of copper-nickel alloys at temperatures above the liquidus temperature dissolve to form a single liquid solution. All compositions of the alloys at temperatures below the solidus line form a single solid solution. Only at temperatures between the liquidus and solidus are two phases present —this is the temperature range of the "mushy" zone in a solidifying casting.

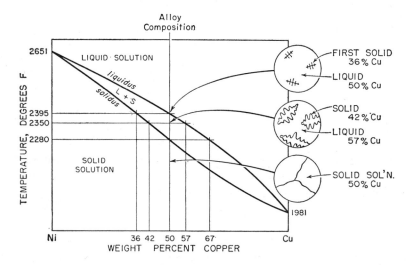

Fig. 4.5. Phase diagram for the nickel-copper alloy system. Microstructures shown are obtained by very slow solidification.

With the aid of the phase diagram, the solidification of a copper-nickel alloy may now be examined. A 50–50 alloy will be chosen; solidification will be considered to be very slow. If such an alloy is melted and poured (with suitable superheat) solidification will begin at 2395°F (1313°C). Nucleation of a solid solution of nickel and copper will take place, and dendrites will begin to grow; the composition of the initial solid will be 36 per cent copper, balance nickel, as shown in Fig. 4.5.

As the alloy cools, the solid grains continue to grow. Since the initial solid contains less copper than the liquid (36 per cent as compared to 50 per cent) the liquid becomes gradually richer in copper. At 2350°F (1288°C), the liquid has increased in copper content to 57 per cent, with the result the solid which is freezing increases in composition to 42 per cent copper. Diffusion of copper in the solid eliminates concentration gradients, and the frozen portion of the melt is a solid solution containing 42 per cent copper (at 2350°F, 1288°C).

Finally, at the solidus (2280°F, 1249°C) the last copper liquid, containing 67 per cent copper, freezes to a solid containing 50 per cent copper. The excess copper atoms from the liquid diffuse into the metal to make the whole solid alloy a homogeneous solid solution of 50 per cent nickel, 50 per cent copper.

Unfortunately (usually) for the metallurgist and the heat treater, solidification in castings is seldom slow enough to attain the ideal diffusion in the solid described above. Such ideal solidification is called *equilibrium solidification*. Most frequently, solidification of castings occurs under *non-equilibrium* conditions, in which lack of complete diffusion in the solid state results in *coring*. In the previous example, the first solid to form contained 36 per cent copper, and the last, 50 per cent. If diffusion had not occurred in the solid, the outer surface of the dendrites would have had a considerably higher copper content than the center; the dendrite would have been "cored" as shown in Fig. 4.6.

Coring in the solidification process is extremely common in cast alloys. It not only alters the final casting microstructure as described above; it also can affect the temperature range over which alloys freeze. For example, in the case of the 50–50 copper-nickel alloy, the continual segregation of copper into the liquid metal eventually results in pools of liquid which contain substantially more than 50 per cent copper. These pools have a lower final

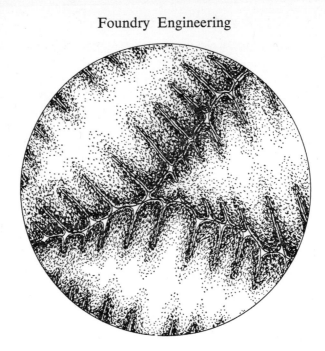

FIG. 4.6. Coring in a nickel-copper cast alloy. Darker areas toward the dendrite surfaces represent increasing copper content.

solidification temperature than that predicted by the phase diagram for the 50–50 alloy, and, hence, solidification of the cored alloy is not complete until a temperature somewhat below that of the "equilibrium" solidus is reached.

In most cases, coring does not appear to be particularly deleterious to the mechanical properties or serviceability of cast parts. If necessary, however, it can be eliminated by reheating the alloy to allow diffusion (homogenization) to take place. Reheating during heat treatment serves the same purpose as casting slowly; the atoms are able to move about (diffuse) more easily and, with enough time at temperature, the structure becomes homogeneous and coring is eliminated.

4.6   Eutectic alloys

Metals that are not mutually soluble in all proportions combine as *mechanical mixtures* to form alloys. Alloys of this type are made up of two chemically (and sometimes structurally) different types of crystals, mechanically mixed on a very fine scale. Some metals are completely insoluble in each other in the solid states; an example of these is tin and cadmium. Alloys of tin and cadmium are made up of very fine crystals of pure tin and pure cadmium. More frequently, however, alloys possess *limited* solid solubility. Such is the case in aluminum-copper alloys. Between 5.65 and 52.5 per cent copper (balance aluminum), these alloys are made up of two different kinds of crystals: one a solid solution containing a high percentage of aluminum, and the other a different solid solution containing a high percentage of copper.

Alloys that freeze as mechanical mixtures usually do so by a *eutectic reaction*. The slow (equilibrium) solidification of an aluminum-10 per cent copper alloy will be described to illustrate the eutectic mode of freezing. The aluminum-rich end of the aluminum-copper phase diagram is shown in Fig. 4.7. An aluminum alloy containing 10 per cent copper, when poured in a sand mold begins to freeze at 1160°F (627°C). The first solid

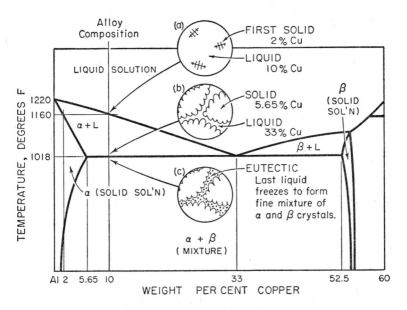

FIG. 4.7. Phase diagram of the aluminum-rich end of the aluminum-copper phase diagram. Microstructures shown are obtained by very slow solidification.

to form consists of primary alpha solid solution dendrites containing 2 per cent copper (98 per cent aluminum), as sketched in
Fig. 4.7a. As the casting cools, solidification continues as in solid
solution alloys. The solid grows and gradually increases in copper
content; at the same time, the remaining liquid also becomes enriched in copper. At just above 1018°F (548°C), the casting
consists of primary aluminum-rich dendrites containing 5.65 per
cent dissolved copper, and a small amount of liquid containing
33 per cent copper (Fig. 4.7b). Now the alpha solid solution
has reached its maximum solubility for copper; no more copper
atoms are able to force their way into the aluminum lattice, and
further solidification can occur only by new, *copper-rich* crystals
solidifying. The temperature at which this occurs is the *eutectic*
temperature (in this case 1018°F, 548°C).

The copper-rich crystals which freeze are beta crystals of 52.5
per cent copper. Simultaneously with their formation, more aluminum-rich crystals are able to form, with the result that at the
eutectic temperature a *mechanical mixture* of fine copper-rich
crystals and fine aluminum-rich crystals freeze to form an aggregate known as *eutectic*. The structure is sketched in Fig. 4.7c.
Because the eutectic freezes at a single temperature, a "hold,"
much like that of pure metal is observed on a cooling curve of the
alloy (Fig. 4.8).

The solidification of an aluminum 10 per cent copper alloy has
been considered in detail to illustrate the eutectic mode of freezing;
other aluminum-copper alloys solidify in similar fashion. For
example, all aluminum-copper alloys that contain between 5.65
per cent and 33 per cent copper have a final structure made up
of primary dendrites containing 5.65 per cent copper and a eutectic
mixture containing 33 per cent copper. The relative amounts of
primary dendrites and eutectic are determined by the alloy composition.

Under conditions of more rapid solidification, such as is obtained in ordinary sand or chilled molds, the primary dendrites
which form during freezing exhibit *coring;* the composition of the
center of the dendrites can be as low as 1 or 2 per cent copper
while the surfaces of dendrites will be 5.65 per cent copper.
Coring in aluminum alloys containing between 5.65 per cent and
33 per cent copper results in an increase in the amount of eutectic
obtained in the structure but does not lower the final solidification

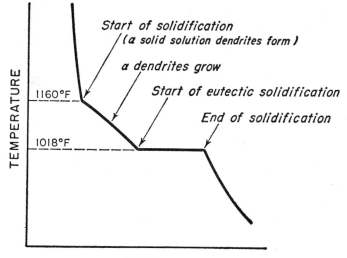

FIG. 4.8.   Cooling curve for an aluminum–10 per cent copper alloy.

temperature.   In the case of aluminum alloys containing *less* than 5.65 per cent copper, coring has a more pronounced effect.   Under conditions of very slow cooling, such alloys would freeze like the solid solution alloys described earlier (with no *eutectic* in the final structure).   However, because of coring, the liquid becomes enriched in copper and can become so enriched that eutectic forms during solidification.   In practice, cast aluminum alloys containing as little as 0.5 per cent copper usually exhibit some eutectic in their final structure; these alloys are not completely frozen at the temperature predicted by the phase diagram but become so only at the eutectic temperature.

The importance of the eutectic mode of solidification to foundry metallurgy will become more evident in later chapters.   Nearly all simple aluminum-base and magnesium-base alloys freeze in this manner, and the most widely used cast metal of all, cast iron, exhibits eutectic freezing.   In some alloys (such as some of the high-strength light-metal-base alloys) the eutectic structure is brittle and damaging to mechanical properties.   In these cases, the eutectic is dissolved by *heat treatment*.   For some purposes (such as wear resistance) certain eutectic structures are desirable,

and no attempt is made to alter them. In gray cast iron, one of
the constituents of the eutectic mixture is graphite which actually
occupies more volume as a solid than when it is dissolved in the
liquid iron. Properly controlled, the expansion of graphite during
the late stages of eutectic solidification can be employed to counterbalance metal "shrinkage" during freezing.

4.7   Intermetallic compounds

Some metals in certain percentages tend to form *compounds*
(*intermetallic compounds*) with one another. Alloys containing

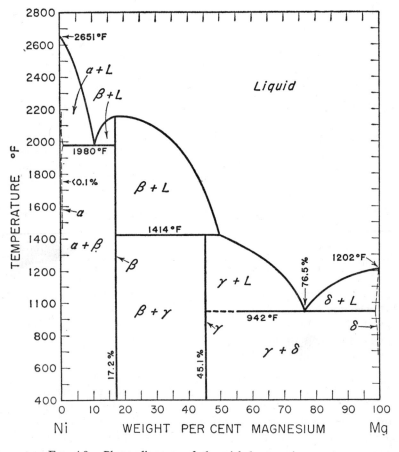

Fig. 4.9. Phase diagram of the nickel-magnesium system.

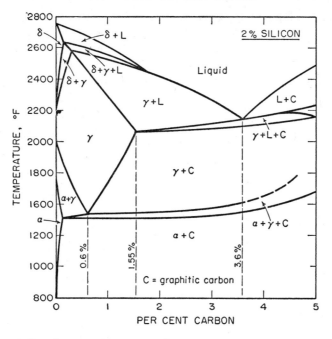

Fig. 4.10. Constant-silicon section of the ternary iron-carbon-silicon diagram.

large amounts of such compounds are technically unimportant as they are for the most part hard, brittle, and lacking in strength and ductility. However, such compounds appear in *small amounts* in many engineering alloys.

Some typical compounds are shown in Fig. 4.9, a phase diagram of the nickel-magnesium system. One, the beta alloy, breaks down on heating to form a liquid; the other (gamma) breaks down on heating to form a liquid plus a new solid. This latter reaction is termed a *peritectic* reaction. It may also occur in the heating of some solid solutions (as in the heating of austenitic iron, Fig. 4.14).

4.8   Phase diagrams

Several phase diagrams for alloys of two metals have already been presented. Many others are available in the literature.[1,2] These diagrams are used by the metallurgist and foundryman in many ways. They indicate the phases present (at equilibrium) in a given alloy at any temperature, and from this information the metallurgist can often deduce what physical and mechanical properties the alloy is likely to possess. He can also determine if heat treating is likely to improve properties of metals and what heat treatments may be useful in improving these properties. Phase diagrams indicate the solidification temperatures of alloys and so are useful in estimating the required pouring temperature for casting a given alloy. They also indicate the range of temperatures over which a casting will solidify; this range of solidification has an important bearing on many foundry characteristics, particularly shrinkage, hot tearing, and fluidity.

The phase diagrams that have been presented here have dealt with alloys of two metals. Most commercial foundry alloys contain three or more metals. The construction of phase diagrams to represent these alloys is difficult at best, and interpretation of such diagrams is time-consuming. In nearly all cases it is adequate to use a "pseudo-binary" diagram to obtain an understanding of the nature of these alloys. Such a phase diagram is shown in Fig. 4.10 for an iron–2 per cent silicon alloy containing varying amounts of carbon.

4.9   Solid state transformations

The transformations that have been discussed thus far have dealt with changes that occur in alloys when they cool from the liquid to the solid state. Changes can also occur when the metal is completely solid. These changes form the basis for all metal heat-treating operations; they are discussed more extensively in Chapter 14, and are only outlined here.

The simplest solid state change is a *composition change* in an alloy phase. An example of this is elimination of *cored* structures by heat treating. It was pointed out that nickel-copper and aluminum-copper alloys possess marked concentration gradients within dendrites when they are solidified at normal rates. If these alloys are heated for prolonged times at elevated temperatures (below the alloy melting temperature), *diffusion* will remove the concentration gradients.

A second type of solid state reaction is a *phase change* which takes place by precipitation of new grains over a range of temperature. Consider the equilibrium cooling of an aluminum-4.5 per cent copper alloy. The alloy, according to the phase diagram of Fig. 4.7 is a homogeneous solid solution at 1000°F. At room temperature, however, the alloy consists of two phases, the original solid solution plus some new crystals of high copper content; this solid state change is the basis for age hardening.

Other changes in the solid state may occur at a single temperature. One such change is a eutectic-like reaction, but where one solid decomposes (on cooling) to form, by diffusion, two different types of crystals. Such a reaction is called a *eutectoid* and accounts for many of the valuable engineering properties of steel. The eutectoid point in steel occurs at 0.80 per cent carbon, and at 1333°F, 723°C (Fig. 4.14). A quite different kind of reaction occurs in certain crystal structures which transform not by diffusion but by mechanical movement, or *shear*. The most notable example of this is *martensite* formation in steel; the eutectoid and martensitic transformations in steel are discussed in detail in Chapter 14.

PART II. METALLURGY OF COMMERCIAL NON-FERROUS CASTING ALLOYS

4.10 Introduction

The binary non-ferrous alloy diagrams illustrated in this chapter are useful guides for understanding the properties of these alloys and their foundry behavior. Like cast iron and steel, most commercial cast non-ferrous alloys are not simple binary metals; they often contain two or three other elements in small amounts which must be controlled to obtain desirable physical and mechanical properties. Where more than a few per cent of a third component is involved, ternary alloy diagrams can usually be found in specialized books on the subject and in handbooks (such as the *Metals Handbook,* A.S.M.).

Most of the popular non-ferrous metals can be readily cast to shape by the usual processes. They can also be heat-treated to homogenize composition and structure or to develop special properties. For alloys susceptible to precipitation hardening, high-temperature solutionizing anneals are used to bring some phase or constituent into solid solution. The part is then cooled rapidly and given a low-temperature hardening treatment. Heat treatments for cast metals are discussed in Chapter 14.

4.11   Aluminum alloys

Aluminum and magnesium and their alloys are useful where weight saving is important. Aluminum and its alloys are chosen also for their relatively good strength, corrosion resistance, conductivity, reflectivity, ease of casting, and relative economy. Aluminum alloys have a strength-to-weight ratio approximately equivalent to cast steels. As compared to mild steel of the same load-carrying capacity, aluminum alloy castings are one-third as heavy. The strength of pure aluminum (99.95 per cent) is too low to be useful structurally, but it can be strengthened by adding copper (to 10 per cent), silicon (to 14 per cent), magnesium (to 10 per cent), or zinc (to 20 per cent). Most aluminum casting alloys contain one or another of these elements as a major alloying constituent. Smaller amounts of other elements may also be present; in addition to those already mentioned, elements such as chromium, titanium, nickel, sodium, and tin are used in alloys for special purposes. Compositions and important characteristics of commercially cast aluminum alloys are listed in several of the references cited at the end of this chapter. These alloys are cast by any of the usual casting methods including sand, die, permanent mold, and plaster casting processes.

4.12   Magnesium alloys

Magnesium, about two-thirds the weight of aluminum and one-fourth that of steel, is the lightest of the commercial metals. For engineering applications, the cast metal is usually alloyed with aluminum (4–10 per cent), zinc (0–3 per cent), and manganese (about 0.2 per cent). More recently developed alloys contain small amounts of zirconium, thorium, or rare earths. Most of these alloys are susceptible to age hardening. They are given a long solutionizing heat treatment at about 700°F (375°C), followed by aging at about 350°F (180°C). The first treatment improves tensile strength and ductility, and the second increases hardness and yield strength with little increase of tensile strength and some loss of ductility. References at the end of this chapter list the exact compositions and characteristics of cast magnesium alloys.

Alloys of magnesium can be cast in specially treated sand molds or in permanent molds and can be pressure-cast in cold dies. Pressure castings are not usually heat-treated, but sand and permanent mold castings are.

4.13 Copper-zinc alloys

Cast *brasses* are alloys of copper and zinc, usually with other agents added; Fig. 4.11 illustrates the copper-zinc phase diagram. The three principal groups of cast brass are *red brass* (2–8 per cent zinc), *semi-red brass* (8–17 per cent zinc), and *yellow brass* (over 17 per cent zinc). In addition to zinc, brasses also contain 2 to 6 per cent tin as a hardener and 1 to 8 per cent lead to improve machinability. High-strength yellow brasses (*manganese bronzes*) are copper-zinc alloys with over 2 per cent total of hardening additions such as aluminum, manganese, silicon, iron, and nickel.

Leaded red brasses are the most popular of the cast copper-base alloys. They are used for such applications as plumbing goods and steam valves where moderate corrosion resistance, good machinability, moderate strength and ductility, and good castability are required. The higher zinc contents of the *semi-red* and *yellow brasses* result in cheaper alloys of lower mechanical properties; they are used mainly in the plumbing field. *Manganese bronzes* possess good corrosion resistance and high strength; they are specified for such applications as marine propellers, gears, and bearing races. Manganese bronzes are, however, more difficult to cast and to machine than other brasses.

4.14 Copper-tin alloys

The copper-rich end of the copper-tin phase diagram is given in Fig. 4.12. Cast copper-tin alloys (*tin bronzes*) contain 5 to 10 per cent tin with 0 to 4 per cent zinc. Lead may be present in quantities of 1 to 10 per cent or more for machinability and to improve casting characteristics. Owing to severe *coring* in tin-bronze alloys, the delta phase (Fig. 4.12) tends to precipitate in alloys containing over about 10 per cent tin. The delta phase embrittles the cast metal and so compositions over about 9 to 10 per cent tin are avoided. Tin bronzes have excellent corrosion resistance and good properties at elevated temperatures and, with lead, make excellent bearings for use at moderate speeds.

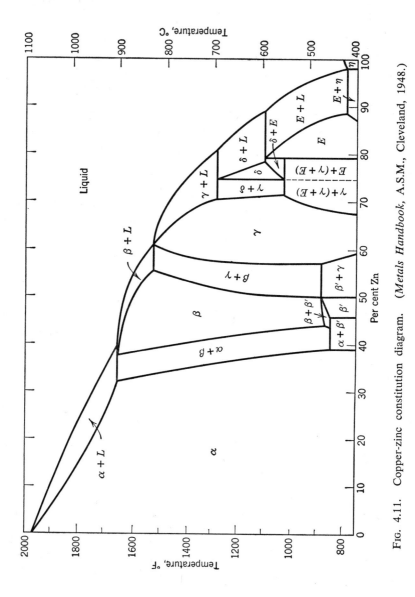

Fig. 4.11. Copper-zinc constitution diagram. (*Metals Handbook*, A.S.M., Cleveland, 1948.)

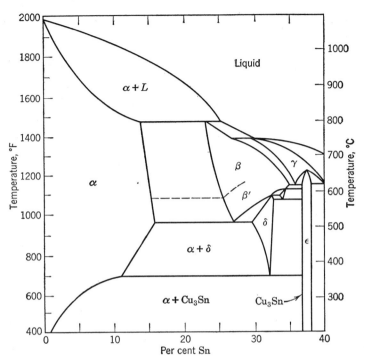

Fig. 4.12. Copper-rich portion of the copper-tin constitution diagram. (Brick and Phillips, *Structure and Properties of Alloys*, McGraw-Hill, 1949.)

4.15 Aluminum bronzes

Commercial cast *aluminum bronzes* usually contain between 4 and 13.5 per cent aluminum, 1 to 4 per cent iron, and some nickel and manganese. They are heat-treated and quenched from 1600 to 1700°F (870–925°C) to give an acicular structure resembling martensite. This is followed by tempering at 900 to 1300°F (480–700°C). The heat treatment increases strength and hardness at some expense of ductility.

Although aluminum bronzes are more difficult to cast than brasses and tin bronzes, they exhibit higher strength, hardness, and resistance to fatigue and shock. They are useful for gears, propellers, pump parts, high-strength bearings, and fans. They also possess higher hardness at elevated temperatures than brasses and tin bronzes. Their color resembles that of 10-carat gold.

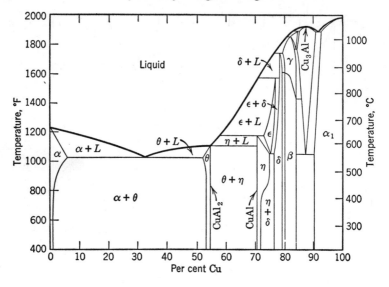

Fig. 4.13. Aluminum-copper constitution diagram. (*Metals Handbook,* A.S.M., Cleveland, 1948.)

The copper-rich end of the copper-aluminum diagram is given in Fig. 4.13. The maximum solubility of aluminum in the alpha solid solution at 1060°F (570°C) is 9.5 per cent. The beta phase, stable at elevated temperatures, undergoes eutectoid transformation at 1050°F (565°C); this transformation makes possible the improvement in mechanical properties obtained by heat treatment.

4.16 Silicon bronze

Most of the commercial cast silicon bronzes lie in the solid solution range. The alloys combine very good resistance to sea water, alkalies, and acids, with good mechanical properties. Besides containing about 1 to 5 per cent silicon, they usually contain smaller amounts of zinc, tin, iron, and up to 0.5 per cent lead; lead is added to improve machinability.

4.17 Zinc alloys

Zinc is most extensively used in castings for alloying with copper to form brass. In recent years, about 20 per cent of the zinc produced has been used for die-casting alloys. These are made of special high-purity zinc and contain about 4 per cent aluminum, up to 2.5 per cent copper, and 0.1 per cent magnesium, with minor impurities. They are commercially known as *Zamak* alloys. The low melting point of these alloys, about 750°F (400°C), facilitates die casting. Such castings of zinc alloys can be plated or otherwise finished to improve appearance and corrosion resistance. The tensile strength of zinc-base die castings may reach 50,000 psi, and their elongation sometimes is as high as 15 per cent. In the automotive, camera, business-machine, and household-appliance fields, these castings find extensive application.

4.18   Nickel alloys

Over 60 per cent of the nickel produced is alloyed with cast iron and steel, and about 15–20 per cent with copper. The 70 nickel, 30 copper alloy, known as *Monel,* is used in wrought and cast form for its corrosion resistance and high strength. Additions of silicon and other elements to Monel make the alloy age hardenable and more wear resistant. The 80 per cent nickel, 20 per cent chromium alloy, known as *Nichrome,* is used for electrical resistance heaters. An 80 per cent nickel, 14 per cent chromium, 6 per cent iron alloy, known as *Inconel,* is used extensively in cast and wrought form. Inconel combines oxidation resistance with high strength at elevated temperatures. An alloy with a low coefficient of expansion known as *Invar* contains 36 per cent nickel, balance iron. *Elinvar* has 36 per cent nickel, 4–5 per cent chromium, and 1–3 per cent wolfram. It has a nearly constant modulus of elasticity over a useful range of temperatures.

PART III.   METALLURGY OF CAST IRON AND CAST STEEL

4.19   Introduction to iron-carbon alloys

In this section, the fundamentals of iron-carbon alloys are to be considered in simple engineering fashion. The effect of heat treating and of alloying agents other than carbon are discussed in Chapter 14. Detailed specifications of chemistry and properties of ferrous casting alloys are not considered, but these can be readily found in the references at the end of this chapter.

The effect of carbon on the structure and properties of iron seems much out of proportion to the amount involved. Less than

0.30 per cent carbon in the iron lattice is sufficient, when properly manipulated, to raise the tensile strength of otherwise pure iron from 40,000 to 130,000 psi. Alloys of iron and carbon, with other elements added for special purposes, make up the important classes

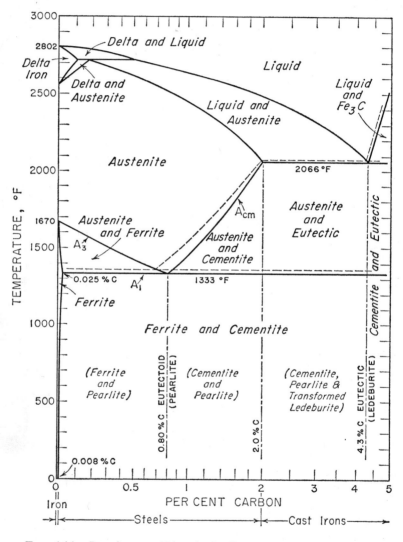

FIG. 4.14. Iron–iron carbide phase diagram. Iron-graphite system shown by dashed lines.

of ferrous metals known as steels and cast irons. These alloys, particularly steels, are amenable to heat treatment, and a wide range of properties can be imparted to them by varying the cycles of heating and cooling.

A phase diagram for iron-carbon alloys is shown in Fig. 4.14. Below 1670°F (910°C), pure iron exists in the body-centered cubic form; it is known as *ferritic iron* or *ferrite*. Ferrite has relatively little solubility for carbon, being able to dissolve a maximum of only about 0.025 per cent. At temperatures between 1670°F (910°C) and 2540°F (1393°C), pure iron assumes a face-centered cubic lattice, and in this form can dissolve up to about 2 per cent carbon. The face-centered cubic form of iron or iron alloys is termed *austenite*.

When the solubility of iron for carbon is exceeded, the carbon precipitates in one of two forms, either as the metastable Fe_3C compound (*cementite*), or as elemental *graphite*. In low-carbon alloys, or higher-carbon alloys cooled at a rapid rate, the carbon always precipitates as cementite (Fe_3C). The phase diagram of Fig. 4.14 is drawn to show the products of the $Fe–Fe_3C$ reactions. At higher carbon contents, cooling rate may be sufficiently slow that the true "equilibrium" graphite will precipitate. The phase diagram for iron-graphite reactions is nearly identical with that for iron-cementite reactions, differing only slightly at the dotted lines shown in Fig. 4.14. Normally cementite rather than graphite precipitates at compositions less than about 2 per cent carbon; this is the composition range of ordinary commercial steels. Above about 2 per cent carbon, either graphite or cementite may precipitate, depending on cooling rate and other factors; the composition range of most cast irons varies from about 2 to 5 per cent carbon. In the following paragraphs the microstructures obtained from the relatively slow solidification of steel are considered; the structures obtained by cooling cast iron at very slow to very rapid rates are then discussed. More detailed discussion of the effect of cooling rates on structure and properties of ferrous materials may be found in Chapter 14.

4.20   Cast plain carbon steels

Cast plain carbon steels of up to about 0.5 per cent carbon solidify by a *peritectic reaction* to form the single-phase solid solution *austenite*. The high diffusion rate of carbon apparently results in very little coring during solidification. Steels of over 0.5 per cent carbon solidify directly to the austenite, over a temperature range. The microstructure of steels in the austenite region is sketched in Fig. 4.15a.

Steels containing between 0.025 per cent carbon and 0.80 per cent carbon cool through a two-phase region where austenite and ferrite coexist. In this range, ferrite precipitates at the austenite grain boundaries and partially consumes the original austenite grains (Fig. 4.15b). At 1333°F (723°C) a eutectoid trans-

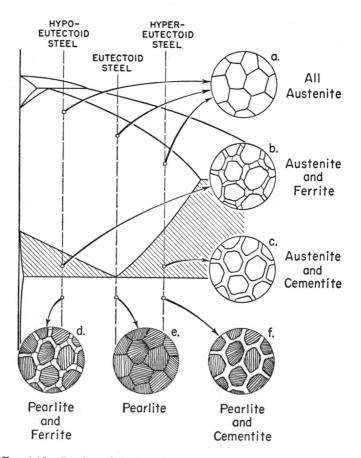

FIG. 4.15. Portion of the iron–iron carbide phase diagram. Microstructures shown are obtained by very slow solidification and cooling.

formation takes place whereby the austenite remaining in the steel transforms directly to a fine, platelike aggregate of ferrite plus cementite, known as *pearlite* (Fig. 4.15d). The eutectoid reaction is similar to eutectic reactions, except that all reaction products are solid in the case of the eutectoid.

Steels of 0.80 per cent carbon transform to an entirely *eutectoid* (*pearlite*) microstructure. They are known as *eutectoid* steels (Fig. 4.15e). Steels containing less than this amount of carbon are termed *hypoeutectoid,* and steels containing greater amounts of carbon are *hypereutectoid* steels.

Hypereutectoid steels begin to transform (on cooling from the austenite region) at the A_{cm} line (Fig. 4.14), with cementite forming at the grain boundaries (Fig. 4.15c). At the eutectoid temperature the remaining austenite transforms to pearlite, as shown in Fig. 4.15f.

4.21 Cast iron

The metallurgy of cast iron is more complex than its economics and, indeed, is one of the most complex metallurgical systems. We have seen that the iron–iron carbide ($Fe–Fe_3C$) system is *metastable* in iron-carbon alloys and that the true *stable* system is iron-graphite ($Fe–C$). This distinction is mostly of academic interest in steels, since graphite seldom forms, but it is of great practical importance in cast irons. If an iron alloy exceeds about 2 per cent carbon, carbon does not have to nucleate from decomposition of austenite; it can form directly from the melt by a *eutectic* reaction. Cementite (Fe_3C), can still nucleate at the eutectic more readily than graphite, but on sufficiently slow cooling graphite itself is able to form and grow.

We will consider in some detail the solidification of a cast iron of 3.0 per cent carbon content. As an approximation, we will assume that the alloy may be cooled at any rate and that we may still use the phase diagram (Fig. 4.16) to predict its structure. We shall cool:

1. *At a rapid rate,* as in a thin section of a sand casting. As the alloy cools below the liquidus, dendrites of austenite form and grow until the eutectic temperature is reached. At the eutectic, graphite formation is suppressed, but austenite and cementite (Fe_3C) precipitate to form *ledeburite,* a form of eutectic which consists of spheres of austenite embedded in cementite. Ledebur-

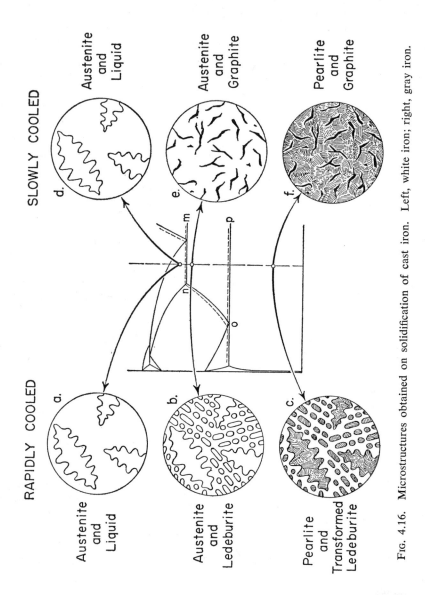

FIG. 4.16. Microstructures obtained on solidification of cast iron. Left, white iron; right, gray iron.

ite forms at the Fe–Fe$_3$C eutectic, solid line *nm*. On further cooling, the cementite grows as the austenite decreases in carbon content along the solid line *no*. At *o* the eutectoid reaction transforms the remaining austenite to pearlite. Microstructures illustrating the solidification and cooling of this alloy are shown in Figs. 4.16a, b, c. At room temperature it is hard and brittle, and is called *white iron* because the surface of a fractured piece of the iron is white and somewhat lustrous. We shall now cool the alloy:

2. *At a slow rate.* In this case, austenite first forms from the melt, but eutectic freezing is now slow enough so that the products of the eutectic reaction are austenite and graphite. The reaction takes place at the dotted line *nm*. The eutectic graphite tends to form as flakes surrounded by eutectic austenite. As cooling continues, the austenite decreases in carbon content along the dotted line *no,* and the graphite flakes grow. At the eutectoid temperature *op,* remaining austenite transforms as before to pearlite. Figures 4.16d, e, and f show typical microstructures of an alloy cooled in this manner; the fracture surface appears dull gray, and the material is known as *gray cast iron* (or *pearlitic gray iron*). Finally, we shall cool:

3. *At a very slow rate.* Phase changes in such an alloy will be exactly like those of case 2 above, except that, at the eutectoid, cooling will be sufficiently slow to permit graphite to precipitate rather than pearlite. Actually, no new graphite flakes will form, but the ones present will increase in size. In this case the final microstructure will be only graphite flakes embedded in a matrix of ferrite; this is termed *ferritic gray iron*. In ordinary castings, cooling is seldom slow enough to obtain this structure, and most ordinary gray iron castings possess the structure of (2) above, graphite embedded in a pearlite matrix. Sometimes the cooling rate of a portion of a casting may be intermediate between (1) and (2); then, a structure containing patches of both white iron and gray iron is obtained—this is known as *mottled* iron.

If we had considered a hypereutectic alloy solidified at various cooling rates, we would have found that by cooling at a moderate or a slow rate the eutectic and eutectoid reactions were the same as those encountered in the consideration of a hypoeutectic alloy. However, the initial solid to form would have been flake graphite according to the stable Fe–C diagram, rather than austenite. Since graphite can nucleate from molten iron comparatively readily, an extremely rapid cooling rate would be required to obtain ledeburite in a hypereutectic alloy. As a consequence, white iron is usually made by solidification of a hypoeutectic alloy at a more easily realized cooling rate.

(a)

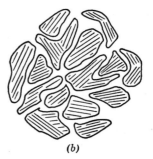
(b)

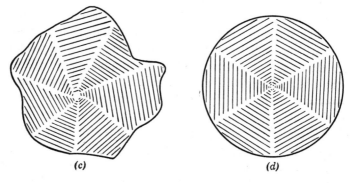
(c)   (d)

FIG. 4.17. Forms of graphite in cast irons.

(a) Basal planes of graphite aligned along major axes of flake in gray iron.

(b) Basal planes of graphite oriented at random in malleable iron containing free manganese sulfide but no free iron sulfide.

(c) Basal planes of graphite oriented perpendicular to radii of nodule in malleable iron containing free iron sulfide.

(d) Basal planes of graphite oriented perpendicular to radii of spherulite in irons treated with magnesium, cerium, or other desulfurizing agent (ductile iron).

4.22 Effects of graphite shape

Graphite may be formed in cast irons as flakes, nodules, or spheroids (Fig. 4.17). Regardless of shape, graphite has very little cohesive strength and reduces the effective metallic cross section of the casting. Graphite reduces tensile strength and ductility most in flake form; the flakes are often interconnected, and their relatively sharp edges are potential areas of low ductility and weakness. The graphite nodule is a more favorable form. The ideal graphite shape from the viewpoint of strength and ductility is spheroidal.

4.23 Iron-silicon-carbon system—commercial gray cast iron

Cast iron is second only to steel in total tonnage produced; in fact, about seven times as much cast iron is made in the United States as cast steel. Typical cast iron has little or no ductility and so cannot be worked; the only way it can be produced is by casting. The tensile strength of gray cast iron ranges from 20,000 to 60,000 psi.

Cast iron has certain unique metallurgical and economic characteristics of interest to the engineer; perhaps the most important is its cheapness. Pig iron from the blast furnace is the starting material for both cast iron and steel. Typical pig iron contains about 3 per cent carbon, 1.5 per cent silicon, and smaller amounts of manganese, sulfur, and phosphorus. The steelmaker must treat this in an expensive open-hearth furnace. He must dilute the pig iron with twice as much or more steel scrap and remove most of the above elements by reacting them at 3000°F (1650°C) under a slag of carefully controlled composition. In contrast, the cast iron foundryman simply charges the pig iron into the top of a vertical shaft furnace called a cupola, which is essentially a miniature blast furnace. Coke and limestone are charged with the pig, and a countercurrent of air burns the coke to furnish heat for melting the iron. Some carbon and silicon are absorbed from the coke and furnace lining, and the slag is self-adjusting. Cast iron of proper composition comes from the cupola spout ready to be poured into castings.

A plain carbon hypoeutectic iron can have a pearlitic matrix (with no massive cementite) if it is cooled slowly enough. In practice, other alloys are always present as impurities or are intentionally added to give a metal of greater strength. To get a

flake graphite structure in an alloyed iron would require imprac-
tically slow cooling. Fortunately, we can speed up graphitization
by adding elements which accelerate the decomposition of cementite. Silicon is the most practical and effective element used for
this purpose commercially; elements that act in this manner are
called "graphitizers." Certain other elements, like chromium, act
to prevent decomposition of the cementite and so are called "carbide stabilizers."

The "pseudo-binary" diagram of the iron-silicon-carbon (2 per
cent silicon) system was described earlier (Fig. 4.10). Comparison of this diagram with that of the plain iron-carbon phase
diagram illustrates the effect of the silicon addition on the solidification of a cast iron. In general, silicon moves the cast iron eutectic
point upward and to the left on the phase diagram, and also results
in a eutectic that does not solidify at a single temperature, but
does so over a narrow range of temperatures.

4.24 Malleable iron

Malleable iron will be discussed at greater length in Chapter 14.
We will here briefly describe only its structure and properties.
Malleable iron is obtained by the heat treatment of white iron so
that the reaction

$$Fe_3C \rightarrow 3Fe + C \quad \text{(graphite)}$$

goes to completion. The heat treatment breaks down the hard
carbide structure of ledeburite and converts it to a matrix of nearly
all ferrite in which iron, *ferritic malleable iron,* has typical mechanical properties of 60,000 psi tensile strength, 40,000 psi yield
strength, and 20 per cent elongation. By other heat-treating
cycles the matrix can be made pearlitic, resulting in a material of
higher strength, and somewhat lower ductility. Metallurgically
the important distinction between malleable and gray iron is that
in malleable iron the graphite is not present in *flakes* which em-
brittle the cast metal, but in *nodules* which are not as deleterious
to strength properties. Hence malleable iron is stronger and more
ductile than comparable gray cast iron.

4.25 Ductile cast iron

A few years ago the discussion of cast irons would have ended
here. One or another variety of flake was the only *as-cast* form

of graphite available to the foundryman. Today we can obtain *ductile* iron without an extensive malleableizing cycle. The key lies in causing the original as-cast graphite to grow as spherulites rather than flakes (Fig. 4.17*d*). To do this, specific agents such as magnesium or cerium are added to the melt. The mechanism of action of these agents is not yet fully understood. To obtain the spherulites, it is necessary to desulfurize the iron almost completely, then add magnesium or cerium and, finally, a "post-inoculant" such as ferrosilicon.

The position of the new "ductile iron" among the engineering metals is not yet clear because it is still such a new development. It is certain to find an important level of use because of its unique characteristics. Typical properties of as-cast ferritic ductile iron are 70,000 psi tensile strength, 50,000 psi yield strength, and 20 per cent elongation; for pearlitic material, 110,000 psi tensile strength, 80,000 psi yield strength, and 6 per cent elongation. Heat treatments for ductile iron are described in Chapter 14.

4.26 Summary

All metals, pure as well as alloyed, are crystalline in nature. They are made up of many *grains* or *crystals,* and within each grain atoms are aligned in ordered arrays. Alloying elements are added to cast metals for a variety of reasons: for example, to improve strength, ductility, corrosion resistance, or casting characteristics. Alloying elements may completely dissolve in the crystal structure of the base metal; then the combination of the two metals (the *alloy*) is termed a *solid solution.* In other cases, the element added is not completely soluble, and the final alloy is a mechanical mixture of two or more types of *grains* of widely differing composition. *Eutectics* are one type of mechanical mixture.

Under conditions of slow cooling, pure metals solidify at a single temperature, but most alloys freeze over a range of temperatures. The range of temperatures over which an alloy freezes determines many of its important casting characteristics. Metals and alloys usually solidify in "pine tree" or *dendritic* fashion; each dendrite grows to form a single grain. In very slowly cooled alloys which exhibit some solid solubility, the composition of each dendrite is uniform, but under the more rapid conditions of solidification, such as are obtained in sand or chilled molds, *coring* takes place. Cored dendrites have a different composition at their center than at their surface.

In Part II of this chapter commercially important nonferrous casting alloys are described from the standpoint of the metallurgist. Details concerning the properties and characteristics of commercial alloys are not considered, but these are readily available in the references listed at the end of this chapter. From a price, strength, and weight standpoint, cast aluminum and magnesium alloys serve well as structural materials; they are readily castable and relatively inexpensive. The selection of copper-base alloys depends on other properties such as ease of machining, wear and corrosion resistance, conductivity, and ease of electroplating. In securing detailed information on specific alloys for design and fabrication purposes, the student is urged to consult the references at the end of the chapter and to make generous use of catalogs published by metal marketing firms. Many suppliers of alloy ingots and of castings maintain technical staffs to advise users and potential users of their products without charge. Another excellent source of information is governmental and society specifications covering cast metals (such as those of the American Society for Testing Materials).

The most widely used foundry alloys are those containing iron and carbon. Steels are alloys of iron that contain less than about 2 per cent carbon; most structural cast steels are of about 0.1 to 0.4 per cent carbon. The composition range of cast irons varies from about 2 to 5 per cent carbon. Other elements are usually added to both steels and cast irons for special purposes.

Below about 2 per cent carbon (in the composition range of cast steels) iron-carbon alloys possess a microstructure made up of iron and iron carbide (Fe_3C). Above 2 per cent carbon, the alloys may freeze either with an iron-iron carbide structure (*white iron*), or with an iron-graphite structure (*gray cast iron*). In gray cast iron, the graphite is present in "flake" form. Mixtures of white and gray iron may be obtained at intermediate cooling rates; this structure is termed *mottled* iron.

On heating, the cementite of white iron decomposes to form nodule-like patches of temper carbon. Iron so treated is *malleable iron*. Suitable treatment of molten cast iron can now be used to produce a structure of graphite spheroids on solidification; such iron, *ductile iron,* is certain to have a bright future.

4.27   Problems

1. The specific heat of liquid iron is 0.146 calorie per gram per degree; that of solid iron is 0.179 calorie per gram per degree; the latent heat of solidification is 63.7 calories per gram. Heat is extracted at the constant rate per unit mass of 10 calories per minute as a casting cools from 2950°F (1621°C), through the freezing point (2800°F, 1538°C), to 2650°F (1454°C). Draw to scale a graph of the cooling curve between 2950°F and 2650°F.

2. Do pure metals always freeze at a single temperature? Why or why not?

3. To what temperature could a very slowly solidified nickel–50 per cent copper alloy be reheated without melting? If solidification were very rapid, at what temperature would melting (on reheating) begin?

4. Silver melts at 1761°F (961°C) and copper at 1981°F (1083°C). A eutectic forms at 1436°F (780°C); its composition is 72 per cent Ag. At the eutectic temperature each metal dissolves 8 per cent of the other; at room temperature, only 1 per cent. Draw the phase diagram. For the alloys containing 5, 20, 28, and 80 per cent Cu, draw time-versus-temperature (cooling) curves, showing the temperatures of initial crystallization and final solidification, and label all changes.

5. For the alloys of Problem 4, sketch the final microstructure of each; label all structures.

6. What are the limitations of alloy phase diagrams from the point of view of predicting (a) temperature where solidification begins, (b) temperature where solidification ends, (c) microstructures at room temperature? In each case, are deviations from what would be predicted by the phase diagram likely to be of importance in ordinary cast alloys?

7. Why are steels containing over 0.80 per cent carbon seldom used for structural purposes?

8. Why is a part made of cast iron usually cheaper than the same part made of steel?

9. How would the structure of an 0.80 per cent carbon steel be affected by cooling rate through the eutectoid temperature?

10. In the reaction $Fe_3C \rightarrow 3Fe + C$, when does the carbon form as flakes? as nodules?

11. What effect does the presence of 2 per cent silicon have on the temperature and composition of the eutectic point in the Fe–C diagram? the eutectoid point? What other important effect does the silicon have?

12. Compare ductile iron with a gray iron of the same composition for (*a*) tensile strength, (*b*) yield strength, (*c*) elongation, (*d*) modulus of elasticity, and (*e*) thermal conductivity. What single factor is primarily responsible for these differences?

13. (*a*) Obtain two samples of aluminum–4.5 per cent copper alloy, one sand cast and the other chill cast. Polish and etch a section of the samples. A very much finer structure will be apparent in the chill-cast sample. Why?

(*b*) Both the above samples possess relatively poor ductility. What heat treatment would you employ to improve the ductility? Perform this heat treatment, and examine the microstructures. Do you think the sand-cast or chill-cast sample is more ductile? Why?

(*c*) What heat treatment might you now employ to improve the yield strength with little loss in ductility? Perform this and again examine the structures.

14. (*a*) Obtain two samples of copper–8 per cent tin alloy, one sand cast and the other chill cast. The sand-cast sample will probably be more ductile than the chill-cast sample. Examine the microstructures of the specimens and explain why.

(*b*) Heat-treat the chill-cast sample in such a way as to improve its ductility. Examine the microstructure to assure that your heat treatment was adequate.

15. Cast steel is heat-treated in various ways to relieve stresses or improve mechanical properties. Heat treatments employed include (*a*) stress relief, (*b*) anneal, (*c*) normalization, (*d*) quench and temper. In the case of a 0.80 per cent carbon steel casting of thin section, what microstructures will result from these heat treatments? How will the tensile strength be affected by the treatments?

16. Obtain a piece of aluminum–4.5 per cent copper (195 alloy) ingot of ordinary commercial purity from a local supply house. The supply house will be able to tell you the exact chemical analysis of the material.

Polish and etch a section of the ingot and, with the aid of the *A.S.M. Metals Handbook* or other suitable reference books, identify all the phases you find present. What impurities do you think are particularly deleterious to mechanical properties?

17. Obtain six gray cast iron specimens, $6 \times \frac{1}{2} \times \frac{1}{4}$ in., and obtain six similar white cast iron specimens (of malleable composition).

(*a*) With the aid of a vise bend a bar of each composition at its center. Measure the maximum angle of bend and compare qualitatively the force required to break the bars.

(*b*) Repeat (*a*) after heat-treating both irons so as to obtain graphite embedded in a ferrite matrix.

(*c*) Repeat (*a*) after heat-treating both irons so as to obtain graphits embedded in a pearlite matrix.

4.28 Reference reading

1. *Metals Handbook,* A.S.M., Cleveland, 1948.

2. Smithells, C. J., *Metals Reference Book,* Interscience, New York, 1955.

3. Williams, R. S., and Homerberg, V. O., *Principles of Metallography,* McGraw-Hill, New York, 1948.

4. Sachs, G., and Van Horn, K., *Practical Metallurgy,* A.S.M., Cleveland, 1940.

5. Buckley, H. E., *Crystal Growth,* John Wiley, New York, 1951.

6. Carpenter, H., and Robertson, J., *Metals,* Oxford University Press, New York, 1939.

7. Samans, C., *Engineering Metals and Their Alloys,* Macmillan, New York, 1949.

8. Hollomon, J. H., and Jaffe, L. D., *Ferrous Metallurgical Design,* John Wiley, New York, 1947.

9. Bullens, D. K., *Steel and Its Heat Treatment,* fifth edition, John Wiley, New York, Volumes I and II, 1948, Volume III, 1949.

10. Murphy, A. J., *Nonferrous Foundry Metallurgy,* McGraw-Hill, New York, 1954.

11. *The Gray Iron Castings Handbook,* Gray Iron Founders' Society, Cleveland, 1958.

12. *Cast Metals Handbook,* fourth edition, A.F.S., Des Plaines, 1957.

13. Briggs, C. W., *The Metallurgy of Steel Castings,* McGraw-Hill, New York, 1946.

14. *Steel Castings Handbook,* S.F.S.A., Cleveland, 1950.

15. *American Malleable Iron,* Malleable Founders' Society, Cleveland, 1944.

CHAPTER 5

Solidification of Castings

PART I. STRUCTURE AND SHRINKAGE OF CAST METALS

5.1 Introduction

Of all the engineering phases involved in manufacturing a quality casting, the least appreciated and most important is the mechanism by which metal freezes. The techniques for melting and handling the metal and preparing the mold are fairly well understood, and are subject to constant, positive control in any good shop. All too often, however, as soon as the metal fills the mold, control ends, and solidification proceeds according to the whims of nature; an element of mystery enters the metals casting process.

Until the advent of modern research and inspection techniques, the solidification of metals was, indeed, a mystery. Even today the many phenomena that influence solidification are not completely understood. Enough is known, however, to permit an *engineering understanding* of the way in which castings freeze. This chapter will treat the basic theory of solidification of metals and alloys from the mechanistic, engineering approach; subsequent chapters will utilize the information in a practical approach to such foundry problems as gating, risering, and chilling.

5.2 Volumetric shrinkage in castings

When metals or alloys solidify and cool, they always contract in volume (with the notable exceptions of bismuth and some of its alloys, and gray cast iron). Cast metals shrink in three distinctly different steps; they undergo *liquid contraction* as they cool from the pouring temperature to the solidification temperature, *solidification contraction* as they freeze, and *solid contraction* as the solid casting cools to room temperature.

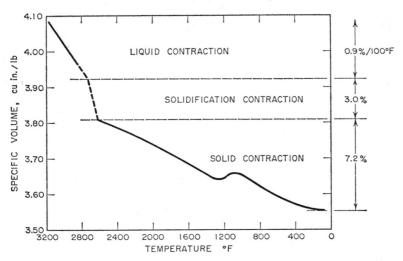

FIG. 5.1. The change in specific volume of solidifying and cooling steel.

Figure 5.1 is a graph of the approximate shrinkages occurring as a mass of molten low-carbon steel cools and freezes. Figure 5.2 is a schematic interpretation of this same information, designed to show the effects of liquid, solidification, and solid contraction on the volume and external shape of a casting. The metal in the imaginary flask, or mold (which is cooled so slowly that no metal freezes to the sidewalls) sinks from L_0 to L_1 as it cools to the point of solidification; this is the *liquid shrinkage* period when the metal gives up its "superheat" to its surround-

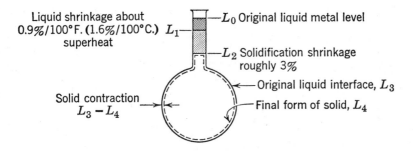

FIG. 5.2. Schematic representation of shrinkage.

ings and cools to the freezing point. For any given mass, cooling slowly and uniformly throughout, there would then be a sudden *solidification shrinkage* and the metal would drop to level L_2. The liquid shrinkage for steel would be about 0.9 per cent per 100°F and solidification shrinkage would be about 3 per cent. These values are approximately correct for most other metals. In a pure metal, the solidification contraction occurs at a single temperature, the melting point. In a nearly pure metal such as the low-carbon steel alloy described, the solidification contraction takes place over a very narrow temperature range. As the solid casting cools to room temperature, it continues to contract and the casting shrinks to a size smaller than the container.

As we will see more clearly later, it is the first two contractions, liquid and solidification, with which we are concerned in risering; of the two, the solidification contraction is by far the most important. The last contraction (solid) results in a casting slightly smaller than the mold cavity, and so patterns must be made slightly larger in size than the desired casting; this is the so-called *"pattern maker's shrinkage."* Solid shrinkage, improperly controlled, also results in internal stresses which may cause *warping* or *hot tearing*.

5.3   Solidification of pure metals

When pure metals, or nearly pure metals such as low-carbon steels, solidify in a mold cavity they do so by forming a solid skin which progresses inward until the entire casting is solid. Figure 5.3a shows the progressive solidification of a low-carbon steel billet. Figure 5.3a is a strictly schematic representation showing that, as the successive layers of solid build up, the liquid level drops.

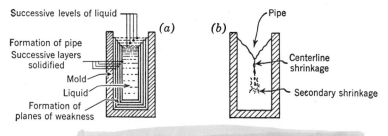

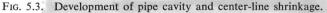

FIG. 5.3.   Development of pipe cavity and center-line shrinkage.

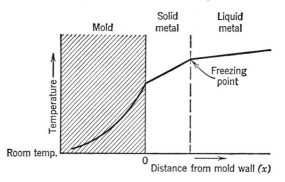

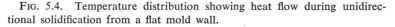

FIG. 5.4. Temperature distribution showing heat flow during unidirectional solidification from a flat mold wall.

The process actually is quite continuous, and the more accurate end result is shown in Fig. 5.3*b*. As the casting solidifies, a *pipe* is formed, primarily from solidification shrinkage. If the walls of the billet are parallel, an elongated pipe may spread nearly to the bottom; this low-density area is called *center-line shrinkage*. The problem of risering, to be discussed in Chapter 6, is basically one of making solidification *directional,* so that all solidification shrinkage occurs in a separate casting appendage, known as a *riser*. The riser is then removed and remelted.

5.4   Rate of solidification of castings

For several reasons it is important to know how rapidly various sections of a casting solidify; e.g., freezing rate has an important bearing on the ability to eliminate solidification shrinkage in a casting (produce *directional solidification*), and on such metallurgical factors as segregation and grain size. In a sand mold, the rate at which a casting freezes is determined primarily by the ability of the mold to accept heat from the casting. This may be seen by considering the very simplified casting of Fig. 5.4   If molten metal is poured against a flat mold wall heated only to room temperature, heat will flow into the mold, and a solid layer of metal will be deposited at the mold-metal interface. As heat continues to flow, the thickness of this layer will increase with time. Temperature distributions in the mold and metal after a given time are shown schematically in Fig. 5.4.   Heat flows from the hot portions to the colder portions of the system according to the relation

$$q/A = -K(\partial T/\partial x)$$

where q/A = rate of heat flow per unit area in Btu/(hr ft^2); K = thermal conductivity in Btu/(hr ft °F); $\partial T/\partial x$ = thermal gradient in units of temperature (T) and distance (x).

The temperature of the boundary between *liquid and solid metal* is, of course, the freezing point of the metal, and heat of solidification is being liberated at this boundary. The rate at which the boundary moves is determined by how rapidly heat of fusion is being removed. Analytical treatment of this transient heat-flow problem leads to useful conclusions.

1. *Solidification rate.* If metal is freezing against a large (flat) mold wall, and heat flow is *normal* to the mold surface, the *thickness* (x) of solid metal deposited will be proportional to the *square root of time* (t), or

$$x = K_1\sqrt{t} \tag{1}$$

This parabolic relationship is shown graphically in Fig. 5.5. Equation 1 indicates that freezing begins at time $t = 0$, and the initial freezing rate is very large since

$$dx/dt = K_1/2\sqrt{t} \tag{2}$$

2. *Solidification time.* If *geometrically similar* castings of different sizes are poured into the same mold material, the *time* for *complete* solidification will be proportional to the *square* of their linear dimensions. This relationship is exact; e.g., a 2-in. steel cube will freeze in one-fourth the time required for a 4-in. steel cube poured at the same temperature.

For comparing castings of different shapes this relationship has been generalized into the form known as *Chvorinov's rule:*

$$\text{Time} = K_2(\text{Volume/Surface area})^2$$

This relationship is not exact but is an excellent engineering approximation.

In actual castings, the rate of solid skin formation is affected by the shape of the casting. Figure 5.6 sketches the appearance

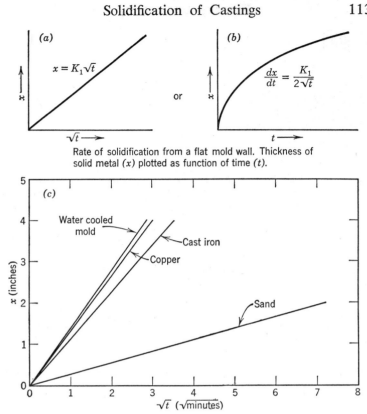

Rate of solidification from a flat mold wall. Thickness of
solid metal (x) plotted as function of time (t).

Solidification of steel from mold walls of different mold materials.

FIG. 5.5.

of some castings which were *bled* at the times shown after pouring. A bled casting is one in which the mold is filled with metal
and allowed to partially solidify; then the mold is inverted and
the liquid remaining is poured out leaving a shell. Note in Fig.
5.6 that the solid skin builds up on flat surfaces roughly as the
square root of time, but that solidification is more rapid at external angles because the heat is more easily dissipated at these
points. Conversely, at internal angles solidification is slower.
Note also the shape of the casting results in *directional solidification* from the thinnest section toward the heavier sections.

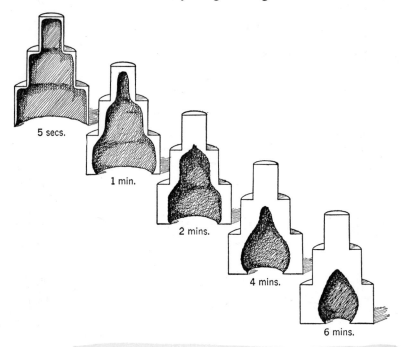

FIG. 5.6. Steel castings "bled" at the times shown; solidified skin remaining.

5.5   Solidification of alloys

We have seen that essentially pure metals freeze by forming a solid skin which grows progressively inward. Only in extremely pure metals, however, is the interface between the solid skin and the liquid metal a smooth wall. When even small amounts of impurities or alloying elements are present, they tend to be rejected by the solidifying metal at the liquid-solid interface. These elements lower the melting point of the liquid adjacent the freezing wall and tend to inhibit further solidification. Freezing then continues by dendrites "reaching" out into the unaffected liquid, with the result that the liquid-solid interface becomes jagged; minute pine-tree-like crystals protrude into the liquid.

In nearly pure metals the dendrites do not protrude far. Figure 5.7a shows schematically the appearance of the liquid-solid interface during the solidification of a low-carbon steel casting.

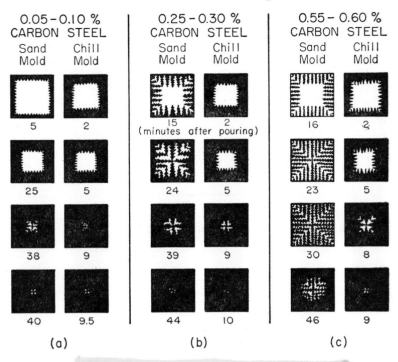

0.05 - 0.10 % CARBON STEEL	0.25 - 0.30 % CARBON STEEL	0.55 - 0.60 % CARBON STEEL
Sand Mold / Chill Mold	Sand Mold / Chill Mold	Sand Mold / Chill Mold

(a)   (b)   (c)

FIG. 5.7.   Solidification of steel in sand and chilled molds.   (After Bishop and Pellini.[2])   Cross sections of 7 in. square ingots are shown at various times during solidification.

Only in the very late stages of solidification do the dendrites reach across the remaining liquid and interlock.   The thermal picture of dendrites reaching into the liquid is sketched in Fig. 5.8.   Figure 5.8a is the extreme upper left-hand corner of the iron-carbon phase diagram.   The vertical line AA' represents the composition of a 0.05 per cent carbon steel.   If the temperatures in a steel casting are measured at a particular time during solidification, a temperature distribution such as that sketched in Fig. 5.8b will be obtained.   Central portions of the casting are above the liquidus temperature, and are, of course, molten.   The outer skin of the casting is below the solidus and is therefore solid.   Between these two zones is a narrow region whose temperature lies between the liquidus and solidus; here, liquid and solid both exist.   In this region, the *mushy zone,* dendrites protrude as shown schemati-

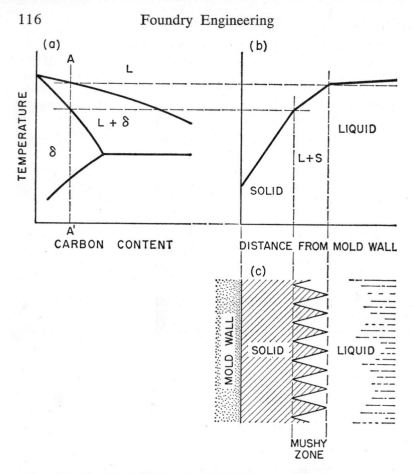

FIG. 5.8. Solidification of an alloy. (*a*) Extreme upper left corner of the iron-carbon phase diagram showing liquidus and solidus of a 0.05 per cent carbon steel (alloy *AA′*). (*b*) Temperature distribution in a solidifying 0.05 per cent carbon steel casting. (*c*) Formation of dendrites in the "mushy zone" (schematic).

cally in Fig. 5.7*c*. The preceding discussion assumes that "coring" (as described in Chapter 4) does not occur. If coring does take place, the picture is modified only slightly by a depressed solidus temperature.

The mushy zone of a low-carbon steel casting is quite narrow; low-carbon steels freeze very much like pure metals. As the carbon content of steel is raised, the mushy zone becomes wider, and solidification becomes less and less like that of a pure metal. In higher-carbon steels, cast in sand, the dendrites may reach into the extreme center of the casting before appreciable solid skin forms (Fig. 5.7c). The reason for this can be seen from Fig. 5.8. The width of the mushy zone of a solidifying casting depends on the solidification range (temperature between liquidus and solidus) of the cast alloy. Alloys that freeze over a large temperature range tend to have a wide mushy zone. The solidification range of steels varies from about 40°F (22°C) for a 0.05 per cent carbon steel to about 140°F (72°C) for a 0.60 per cent carbon steel.

Just as a short solidification range tends to narrow the mushy zone, so steep *temperature gradients* do also. Cast steels tend to have relatively steep temperature gradients during solidification (compared to other cast alloys) because of their high freezing temperatures and low thermal conductivities. The thermal gradients may be made even steeper by *chilling* (casting against a metal mold wall). Figure 5.7 shows schematically the effect of chilling in narrowing the mushy zone of solidifying steel castings.

Most foundry alloys other than steels freeze with mushy zones even wider than that of high-carbon steel. These alloys generally freeze over a fairly wide temperature range, and freeze with flatter temperature gradients than do similar steel castings. Some cast iron, aluminum, and magnesium alloys, and most brasses and bronzes freeze in sand molds without forming any appreciably solid "skin." In aluminum alloys, for example, central portions of a casting are often more than 90 per cent solid before the surface is completely frozen.

5.6   Grain structure of cast metals

When a hot molten metal is poured into a cold mold, the initial rate of heat extraction is very high. The liquid metal near the mold wall surface is often cooled below its freezing point (undercooled), and many fine *equiaxed* dendrites form in this narrow surface zone, termed the *chill zone*. After the initial thermal shock, those dendrites oriented most favorably grow perpendicular to the mold wall to form elongated or *columnar* dendrites. Pure metals, or metals that freeze with a very narrow mushy zone may have only these two types of grains: *chill* and *columnar* (Fig. 5.9a).

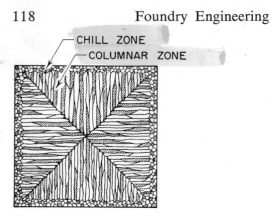

(a)

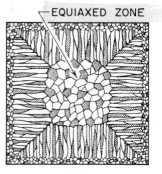

(b)

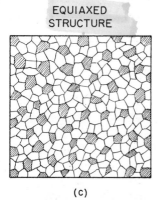

(c)

FIG. 5.9.   Grain structures in cast metals.   (a) Structure characteristic of a pure or nearly pure metal.   (b) Intermediate structure.   (c) Structure characteristic of a metal that freezes with a wide "mushy zone."

$V_2 - V_x$

Metals that freeze with a wider mushy zone tend to *nucleate new grains* during late stages of solidification, in the region where liquid and solid coexist. These new grains are equiaxed, and the region where they appear is termed the *equiaxed zone*. Figure 5.9*b* sketches the cross section of a simple casting of an alloy such as medium-carbon steel which may show all three types of grain structure: i.e., *chill, columnar,* and *equiaxed.*

In cast irons and most non-ferrous alloys, the mushy zone is sufficiently wide that a columnar zone is seldom observed. Rather, fine equiaxed grains tend to form throughout the casting, as shown in Fig. 5.9*c*. The addition of *grain refiners* also tends to reduce any tendency for columnar grains to form; grain refinement is discussed in greater detail in Chapter 10.

5.7 Solidification shrinkage of alloys

In the case of alloys that freeze in a very "mushy" manner, it helps to visualize little pools of liquid metal which tend to be caught between the dendrites and between dendrite arms, late in the freezing process. If liquid channels were not kept open to "feed" the liquid pools, solidification shrinkage would result in small, isolated pores throughout the casting. This can, in fact, occur.

Figure 5.10 illustrates the solidification shrinkage characteristics of a number of foundry metals and alloys. The enlarged portions of these illustrations are *risers* (feed heads, feeders, shrinkage reservoirs, etc.). Risers are discussed extensively in the next chapter. The castings in the top row of sketches show the shrinkage behavior of three different essentially pure metals: aluminum, copper, and steel. The total shrinkage decreases in the order listed, because the volumetric solidification shrinkage decreases in that order (aluminum, 6.6 per cent; copper, 4.9 per cent; steel, about 3 per cent). In the second row are shown two alloys that freeze in a very "mushy" fashion, aluminum–7 per cent silicon, and copper–10 per cent tin bronze. In both castings gross shrinkage is very low, but the casting contains interdendritic shrinkage too fine to see with the unaided eye, except on a cut and polished section through the porous area. The last sketch in the second row shows the shrinkage behavior of another alloy that freezes in a very "mushy" manner, gray cast iron. Here, however, the interdendritic areas are not porous

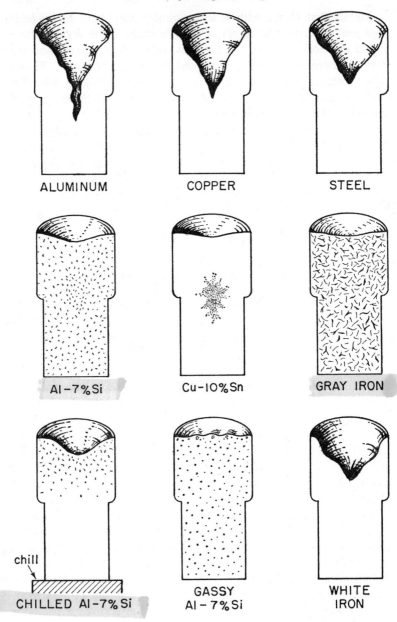

ALUMINUM COPPER STEEL

Al–7%Si Cu–10%Sn GRAY IRON

chill
CHILLED Al–7%Si GASSY Al–7%Si WHITE IRON

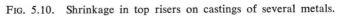

FIG. 5.10. Shrinkage in top risers on castings of several metals.

but are filled with flakes of graphite. Properly controlled, the graphite expansion which occurs in the solidification of gray iron can completely counteract metal shrinkage.

In the bottom row, chilling is shown to minimize microporosity (by narrowing the mushy zone). Dissolved gases which precipitate on solidification, however, accentuate microporosity. The last sketch in the bottom row shows that when cast iron solidifies as white iron it shrinks markedly, because of the absence of graphite formation.

PART II. GASES IN CAST METALS

5.8 General considerations

The effect of gases on cast metals may be fairly simply described. Gases are almost universally more soluble in liquid metals than in solid metals. As metals cool, and particularly as they freeze, dissolved gases tend to be expelled from solution to form *gas holes*. The gas holes may be in the form of large *blowholes,* finer *pinholes,* or still finer *microporosity,* depending on the amount of gas present and on the manner in which the metal or alloy freezes. These gas holes in their various forms can be quite detrimental to mechanical properties, to pressure tightness, and to the surface finish of machined or cast surfaces. Every effort should be made in the foundry to assure that molten metal is free of all dissolved gases before it is cast.

Gases dissolve in metals not as molecules (such as H_2, CO_2, etc.) but as elements (as H, O, C, etc.). On solidification, the elements may recombine to form molecular gases, and, in doing so, form gas holes. Table 5.1 lists elements that dissolve in some common foundry metals (or their alloys) and subsequently com-

TABLE 5.1

SUMMARY OF ELEMENTS THAT DISSOLVE IN VARIOUS METALS, AND
PRECIPITATE ON SOLIDIFICATION TO FORM VOIDS

Metal	Elements Dissolved	Gases Evolved on Solidification
Mg	H	H_2
Al	H	H_2
Cu	H, C, O, S	SO, SO_2, H_2, H_2O, CO, CO_2
Fe	H, O, N, C	H_2O, CO, CO_2, N_2, H_2

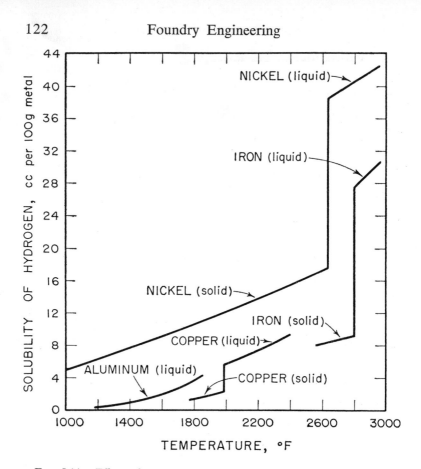

Fig. 5.11. Effect of temperature on solubility of hydrogen in various metals. (After Bever.[8])

bine to form gases. Hydrogen is the most common of the elements (and usually the greatest troublemaker); it is soluble in all metals to a greater or lesser extent. Figure 5.11 indicates the behavior of hydrogen in various metals; it is typical of many other gases. The gas is quite soluble at high temperatures in the liquid alloy. As the temperature decreases, the gas solubility also decreases until the solidification temperature is reached. Here a discontinuous drop in solubility occurs, and it is this drop that accounts for the greater portion of gas porosity in castings.

5.9   Prevention of gas absorption

Gases are absorbed in molten metals (1) in the furnace from the atmosphere and from wet charges, (2) in the furnace or ladles from incompletely dried linings or wet tools, (3) in passing through air from furnace to ladle or from ladle to mold, and (4) in the mold from moist atmosphere or excess moisture or volatile hydrocarbons in mold materials.

Some precautions against gas absorption are (1) melt under slags or protective atmospheres or in a vacuum, (2) reduce chances for absorption of gas by keeping *all* items that must come in contact with the metal dry, (3) melt and pour at temperatures as low as possible to minimize gas solubility, and (4) handle (skim, stir, or reladle) the metal as little as possible.

It is usually impractical to prevent gas absorption entirely. For example, some hydrogen gas is nearly always present in molten metals unless they are melted *in vacuo;* hydrogen reacts with oxygen, usually present as an oxide of the metal or of one of its alloying elements, to form water vapor. It is suspected that water vapor, hydrogen, and oxides of carbon (CO and CO_2) are responsible for most pinholes in castings.

5.10   Gas removal

When it is not possible to prevent gas absorption, the hazards of dissolved gases may be reduced by (1) removing the gas or gases, or (2) removing or isolating one or more of the components of the gas.

One example of the first method of removal depends on the fact that gases are least soluble in cold, solid metal. Thus gases may be removed by a sequence of *melting, freezing,* and *remelting.* Except in special cases, however, this method is too expensive and time-consuming to be used in practice. More common techniques of gas removal rely on the fact that the amount of gas dissolved in a molten metal depends on the pressure of that gas above and around the metal. In the case of diatomic gases such as H_2, O_2, or N_2, the amount dissolved at any particular temperature may be expressed by the equation:

$$V = K\sqrt{P} \tag{3}$$

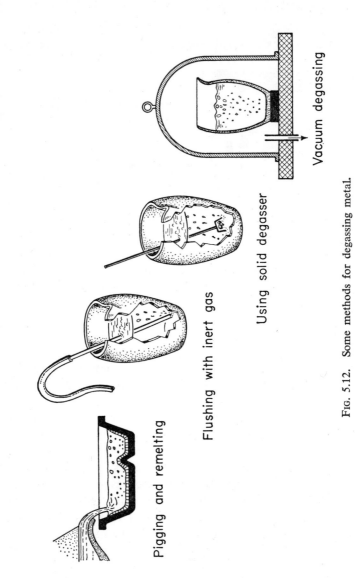

Pigging and remelting

Flushing with inert gas

Using solid degasser

Vacuum degassing

Fig. 5.12.   Some methods for degassing metal.

where V is the volume of diatomic gas dissolved, K is a constant, and P is the pressure of the gas above and around the metal. The equation shows that, if hydrogen is completely removed from the atmosphere around a melt, any hydrogen gas dissolved will tend to escape. Thus melting and pouring under a vacuum is one method, but a very expensive one for obtaining gas-free castings. *Vacuum degassing* is an attempt to combine the advantages of vacuum melting with the economy of air melting. Here the metal is melted in air and then placed in an evacuated chamber (for degassing) before pouring; pouring takes place in the open atmosphere. *Flushing* with an inert gas is the most common method of removing gas employing the principle described above. In this method, the inert gas is bubbled directly through the metal; the pressure of the dissolved gas (usually hydrogen) within each inert gas bubble is zero, and any dissolved gas migrates readily to the bubble where it collects and is flushed from the metal baths.

Inert gases such as chlorine, nitrogen, or argon are used in practice for degassing metals, especially aluminum and magnesium alloys. The gases are simply piped from storage tanks through graphite tubes directly into the metal baths. Sometimes *solid degassers* are also used as a source of the inert gas; here a solid compound (such as C_2Cl_6) held beneath the melt surface decomposes to give the flushing action. In steelmaking practice, the CO from the carbon boil ($FeO + C = CO$) acts as an inert gas to flush hydrogen from the metal. Figure 5.12 illustrates several of the degassing techniques described above.

Gas porosity is also minimized by removing or isolating one of the components of the gas. Thus in steelmaking, after the carbon boil has progressed to the point desired it is *blocked* by adding silicon or aluminum to the bath to tie up one of the reagents of the CO gas ($3FeO + 2Al \rightarrow Al_2O_3 + 3Fe$). Another example is the addition of phosphor-copper to copper-base alloys to tie up dissolved oxygen. The phosphorus combines with dissolved oxygen to form a solid, P_2O_5, which floats out of the bath, and prevents the oxygen from combining with any dissolved hydrogen to form water vapor.

5.11 Summary

The solidification of metals has been discussed in sufficient detail to provide an engineering understanding of the way metals freeze.   Such an understanding is essential before approaching such practical problems as gating, risering, and chilling of castings.

Metal, during freezing and subsequent cooling undergoes three contractions—liquid, solidification, and solid.   Extra liquid metal must be available at all times during freezing to feed the first two of these shrinkages; otherwise cavities will develop.

Pure or nearly pure metals solidify by forming a solid skin which then grows progressively into the casting.   When the mold wall is flat, the thickness of solid skin is proportional to the square root of time.   Solidification at internal angles is somewhat slower, and that at external angles is somewhat more rapid.   In sand castings of simple shapes, the time for complete solidification may be roughly expressed by Chvorinov's rule:

$$\text{Time} = K_2 \, (\text{Volume/Surface area})^2$$

Most alloys freeze over a range of temperatures; during solidification of alloys, a "mushy zone" consisting of solid dendrites intermixed with liquid exists over at least a portion of the casting.   If solidification is not sufficiently *directional,* isolated pools of interdendritic liquid may not be properly fed, with resulting *microshrinkage.*

In castings, zones of *chilled* grains, *columnar* grains, and *equiaxed* grains may be present.   Not all these structures are present in all castings; their presence or absence depends on alloy composition, solidification rate, and other factors.

Gases in metals are always a potential hazard to the foundryman; he must keep melting stock, crucibles, and tools clean and dry, and he must melt metal under slags or a controlled atmosphere; otherwise the hot, molten metal, which readily absorbs gases in direct proportion to its temperature, is unfit for use.   Metals absorb only elemental gases, but reactions such as $H_2O \rightarrow 2H + O$ occur readily when compounds come into contact with hot metal.   Gases may be *tied up* in metal by promoting a reaction to stable, insoluble compounds, or they may be removed by flushing the metal with an inert gas, or by a melting-casting-remelting cycle which purges the gas during the first freezing process.

Gases are soluble in metal in direct proportion to the square root of the pressure of the gas or gases above or about the metal. If a particular metal is sensitive to a certain gas, its dissolution can be prevented by reducing the pressure of the gas concerned to zero; this can be done by melting *in vacuo* or in an atmosphere of some inert gas to exclude the undesirable gas.

Castings of aluminum and magnesium are very susceptible to gas unsoundness; liberal use of chills is an excellent expedient, particularly if alloys of these metals are used, or if ingot composition is not rigidly controlled. Every foundryman should carefully control melting, pouring, and mold atmospheres, melting stock, and temperatures of melting and pouring, and should scrupulously avoid wet tools, ladles, or furnace linings; he should know (to the extent currently possible) what gases are dangerous to his particular metal, how to keep such gases out of his metal, and how to get them out or control them if contamination does develop.

5.12 Problems

1. List the three types of volumetric shrinkage for steel. Describe a method for making shrink-free ingots.

2. Sand castings *A* and *B* are geometrically similar. *B* weighs three times as much as *A; A* freezes in 10 minutes. How long will it take *B* to freeze?

3. A 6-in. magnesium cube freezes in a sand mold in 5 minutes. How long will a magnesium sphere 6 in. in diameter require to solidify completely?

4. Why is solidification retarded at internal angles of a casting? What practical problems do you think this effect might cause?

5. Why is the rate of solidification of a sand casting important?

6. Pure metals freeze by forming a solid skin with a smooth wall between solid and liquid; most alloys do not. Why not?

7. Why is the length of the "mushy zone" in a freezing casting important? What variables affect this length?

8. What are the grain structures that may be present in a metal casting? What variables affect whether or not these structures are present?

9. What types of alloys are most likely to exhibit microshrinkage? List four commercially used alloys of this type.

10. What types of alloys are most likely to exhibit pipe and centerline shrinkage? List four commercially used alloys of this type.

11. What causes positive and negative macrosegregation?

12. In sand-cast bronze valves, chills are frequently used in the region where the valve seat is to be machined. Discuss briefly.

13. Name three ways in which chills may benefit a casting.

14. Briefly discuss the solution of gases in metals. What simple

precautions should be taken in the foundry to minimize gas dissolution in metal?

15. How does Sievert's law apply to the degassing of metals by the use of inert and relatively insoluble gases?

16. Discuss briefly the principles involved in degassing a metal by melting, pigging, and remelting.

17. What methods would you choose for degassing the following metals in a production foundry: (a) aluminum, (b) low-carbon steel, (c) tin bronze? Why?

18. Follow the progress of solidification in an ice-cube tray by "dumping out" individual ice cubes after varying times. Investigate how the shape of the solidification front is affected by insulating the top of the ice-cube tray, and by the corner effects of the tray.

19. Repeat Problem 18 with pure aluminum in a simple sand mold.

20. Investigate solidification in more complex shapes by using water as a model and using the "dump out" technique. Glue pieces of polyethylene together to form the sand mold. To place a "chill" in the mold cut a hole in the polyethylene and glue a penny in place. Sketch the mold design you employ, together with the progress of solidification in the mold.

21. Would you expect Chvorinov's rule to hold for water freezing in a mold comprising a sheet of polystyrene? Does it?

22. Obtain an "ingot mold" of polyethylene; one that is 2 in. in diameter by 2 in. high is satisfactory. Fill the mold with pure water, and add a small amount of dye (or water color); place in a freezer and allow the water to solidify at a relatively slow rate. At various times during freezing, remove and observe the ingot, and note (a) shape of the solidification front, (b) smoothness of the solid-liquid interface, and (c) segregation of the dye.

23. How do you think the results you obtained in Problem 21 would change if you added (a) a pinch of salt? (b) ¼ teaspoon of salt? Do these experiments.

24. If you were to pour a series of cylinders 3 in. in diameter by 4 in. high of pure Al, Al–0.5 per cent Cu, and Al–4.5 per cent Cu:

(a) How would you expect a cooling curve at the center of each ingot to appear?

(b) What would you expect would be the appearance of an etched macrosection from the ingot (in the vertical plane at the center)?

(c) How would the microstructure of the bottom of the ingot appear?

(d) Check your expectations by experiment.

25. Should an aluminum–4.5 per cent copper alloy casting be risered differently from a casting of pure aluminum? Why? Cast some plates $5 \times 10 \times \frac{1}{2}$ in. high, and determine the risering necessary

to make plates of the alloys sound.   Section the plates vertically to determine presence or absence of shrinkage.

5.13   Reference reading

1. Ruddle, R. W., *The Solidification of Castings,* Institute of Metals, London, 1957.

2. Bishop, H. F., and Pellini, W. S., "Solidification of Metals," *Foundry,* **80,** 87, February 1952.

3. *The Solidification of Metals and Alloys,* A.I.M.M.E., New York, 1951.

4. *The Principles of Basic Open Hearth Steel Making,* A.I.M.M.E., New York, 1950.

5. *Metals Handbook,* A.S.M., Cleveland, 1948.

6. Murphy, A. J., *Nonferrous Foundry Metallurgy,* McGraw-Hill, New York, 1954.

7. Adams, C. M., Jr., Flemings, M. C., and Taylor, H. F., "Solidification and Risering of Gray Iron Castings," *Transactions A.F.S.,* **66,** 1958.

8. Bever, M. B., "Gases in Cast Metals," *Iron Age,* **161,** April 22, 1948.

CHAPTER 6

Risering

PART I.—RISERS AND RISERING

6.1 The reason for risering

To develop an appreciation of risers as they are designed and used by foundrymen, it is useful to make an imaginary casting. For example, let this imaginary casting be a cube, $6 \times 6 \times 6$ in., surrounded completely with molding sand and filled instantaneously with molten metal; no gating system will be used, and the metal will be pure iron with no gas or extraneous impurities. The gate is deleted to avoid its effect on thermal conditions. Figure 6.1*a* illustrates the condition of the imaginary experiment.

The metal in the $6 \times 6 \times 6$-in. cavity will begin to freeze at the mold-metal interface, and a so-called *skin* of solid iron will form as heat is extracted by the mold. As the liquid freezes, it shrinks in volume and attempts to create a *partial vacuum* within the casting. This partial vacuum may lead to a shrinkage void which will start to form just under the solidified skin at the top of the cube. As solidification continues, the void will grow as the metal in this location compensates for the solidification shrinkage elsewhere. Most of the heat will flow out the sides and bottom of the cube since the air gap between the liquid metal and solid metal skin at the top will act as an insulator. Solidification will proceed along the isothermal lines as sketched in Fig. 6.1*c*, and the bottom of the shrinkage cavity will occur at the thermal center (hottest part) of the system which will be located near the top center of the casting.

It is possible that some of the solidification shrinkage could be accounted for by external displacement of the casting wall, as shown in Fig. 6.1*d*. In this case, the central cavity would be re-

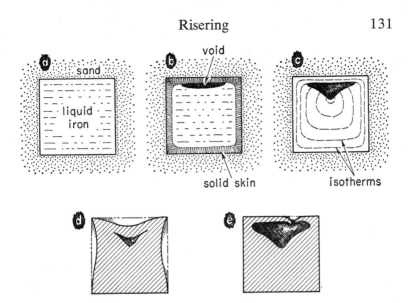

Fig. 6.1. Solidification shrinkage of an iron cube. (*a*) Initial liquid metal. (*b*) Solid skin and formation of shrinkage void. (*c*) Internal shrinkage. (*d*) Internal shrinkage plus dishing. (*e*) Surface puncture.

duced in volume. Displacement of the casting wall develops because of the solidification shrinkage which causes a partial vacuum within the casting. If the difference in pressure between the internal portions of the casting and atmospheric pressure becomes great enough to overcome the strength of the solidified shell, the walls buckle, or collapse ("dish"). Dishing is most severe in metals that have a very weak skin strength, such as aluminum- and copper-base alloys.

It is also possible that the upper skin of the casting would completely collapse (be punctured) early in the freezing process, and the shrinkage cavity would be modified as in Fig. 6.1*e*. Since air would have ready access to the shrinkage hole through the upper shell, any difference in pressure tending to occur would be neutralized, and the external wall would not collapse. We can now make certain deductions:

1. Since shrinkage cavities are not tolerable in castings we must arrange to get rid of them. Risers are used on castings for this purpose. Figure 6.2*a* shows how a riser would accomplish the mission of removing shrinkage from the 6-in. cube in question.

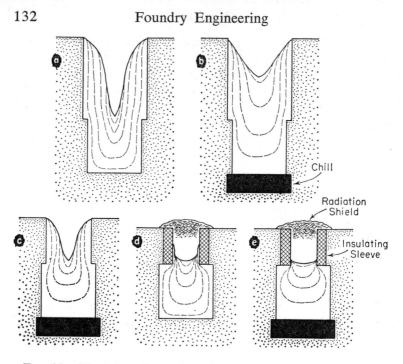

FIG. 6.2. Risering an iron cube. (*a*) Open-top riser. (*b*) Open-top riser plus chill. (*c*) Small open-top riser plus chill. (*d*) Insulated riser. (*e*) Insulated riser plus chill.

2. If a riser should be inadequate, negative pressures would develop at some point in the casting proper, and internal shrinkage cavities would form, or the surface would collapse or puncture.

3. An adequate riser is one that eliminates objectionable solidification defects.

6.2 Requirements of a riser

We have seen the results of inadequate risering, and are now in a position to establish the requirements of an adequate riser. Such a riser must be large enough to freeze after the casting and to contain enough liquid metal to feed the solidification shrinkage of the casting. An adequate riser must also be kept open to the atmosphere and placed in such a location that it maintains a positive pressure of liquid metal on all portions of the casting it is to feed. Stated differently, a sound casting will result only if solidification proceeds directionally from the casting toward a pool of readily available molten metal.

The above conditions might better be considered, perhaps, as "minimum requirements of a riser." There are other requirements, primarily economic in nature, that a riser should meet. For example, a riser should be designed and treated so that it will accomplish its mission with minimum waste metal and should be placed in a position, relative to the portion of casting it is to feed, to provide maximum efficiency. These considerations will be more evident later.

6.3   Riser size and directional solidification

Historically, experience has been the best guide in the risering of castings. It still is. Theoretical analyses of riser requirements for castings of some metals have been developed and are often useful (these are discussed later in this chapter), but most castings are still risered by the "cut and try" method. Let us return to the iron cube of Fig. 6.1, and deduce an adequate size of top riser to completely feed the casting. We might imagine that we would need somewhat more metal in the riser than the casting; and in this case practical experience would prove us right. The "yield" (weight of metal in casting × 100/weight of metal in casting + weight of metal in risers and sprues) would be very low, probably somewhere between 35 and 40 per cent. For every 100 lb of metal poured, we would be able to sell only 35 to 40 lb! This is not a very economical balance, and every effort should be made to increase the ratio of salable to waste metal. The foundryman uses many techniques to reduce the riser size required for a given casting; we will first examine these techniques in a general way with the aid of our 6-in. cube, and later study them in more practical detail.

Figure 6.2a shows the shrinkage cavity existing in the riser of an adequately risered 6-in. cube of pure iron. The freezing isotherms (lines of equal temperature) show the progress of solidification. The last metal to freeze is just at the base of the riser, outside the casting. Yield for this casting has been described as about 35 per cent.

If, now, heat can be extracted more quickly from the casting than from the riser (i.e., solidification made more directional), the large riser may be reduced in size. Figures 6.2b and c show one method of doing this. A large metal *chill* is molded at the base of the cube; the chill extracts heat from the casting much more readily than does the sand mold, and so the *hot spot* (last metal to solidify) tends to move up into the riser, as shown in Fig. 6.2*b*. The large riser employed in the cube of Fig. 6.2*b* is then unnecessary; a much smaller riser is adequate, as shown in Fig. 6.2*c*.

Insulating risers increase casting yield in much the same way as do chills, by promoting directional solidification toward the riser (Fig. 6.2*d*). Of course, the optimum condition and highest yield is obtained by using both a chill and insulation around the riser (Fig. 6.2*e*).

In practice, all these expedients are used in one or another modification for directing solidification. Open risers should always be covered with an insulating or exothermic powder to prevent loss of heat to the atmosphere. Internal or external chills are placed at heavy sections, such as bosses, corners, or adjoining sections, to increase the cooling rate of such regions; insulating sleeves or pads are used on risers and thin sections to retard solidification.

6.4   Riser location and directional solidification

In castings which are not as simple as the cube we have been studying, it is often impossible to obtain directional solidification toward a single riser simply by making the riser larger. Figure 6.3 shows one example of such a casting. No matter how large the riser might be in Fig. 6.3*a*, shrinkage will always occur in the isolated heavy section as shown. Solidification occurs directionally toward *two* hot spots; one of them was fed as shown by the riser, but the other hot spot was unfed. Three solutions are readily apparent. One is shown in Fig. 6.3*b*, where two risers are used, and directional solidification is promoted from the center of the casting toward the two risers. A second solution is shown in Fig. 6.3*c*, where a chill has been used to eliminate the isolated hot spot: i.e., promote directional solidification. Lastly, directional solidification may be promoted by using insulation at the thin section of the casting, Fig. 6.3*d*.

Shrinkage may occur in castings of uniform section owing to lack of directional solidification (even though a very large riser may be used). Figure 6.4*a* shows a "center-line" shrinkage defect which would result in a top-risered steel casting of the general shape shown, regardless of how large a riser might be used. This defect again results from lack of directional solidification toward the riser. Figure 6.4b shows how, late in the solidification process, dendrites "bridge" over and lock a pool of molten metal at the center of the casting. The solution is to improve directionality of solidification. Additional risers could be attached to the side

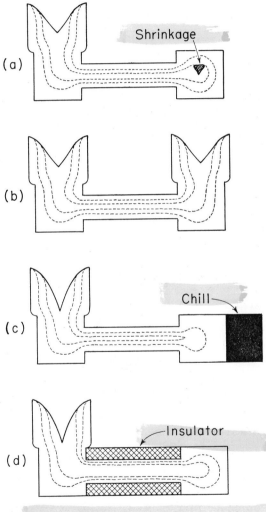

FIG. 6.3. Risering an isolated heavy section.

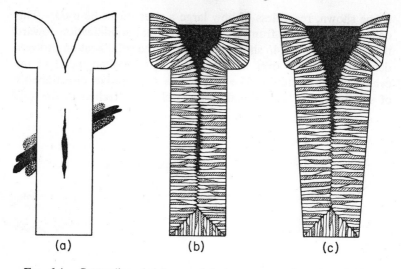

(a)   (b)   (c)

FIG. 6.4.   Center-line shrinkage.   (a) Appearance of sectioned casting.
(b) "Bridging" of dendrites in solidification.   (c) Padding to avoid "bridging."

of the casting to accomplish this, or, often more economically, the
casting could be *tapered,* or *padded* as shown in Fig. 6.4c to promote directional solidification.   *Padding* may be used in conjunction with chills and insulation to obtain optimum directionality
of solidification.

6.5   Atmospheric pressure and risering

Atmospheric pressure is an important and not always well understood factor in the risering of castings.   Atmospheric pressure,
improperly applied, may result in *dishing* or *puncturing* as in Fig.
6.1e.   Properly applied atmospheric pressure can, however, substantially improve riser effectiveness.

A basic rule of risering is that the riser must remain open to
atmospheric pressure throughout solidification.   Figure 6.5a illustrates a casting in which this rule was not followed.   A solid skin
froze over the riser prematurely; a vacuum tended to form within
the casting as it solidified, and the skin punctured at its weakest
point, not in the riser but at the hot spot of the casting.   By insulating the top of the riser, as shown in Fig. 6.5b, skin formation

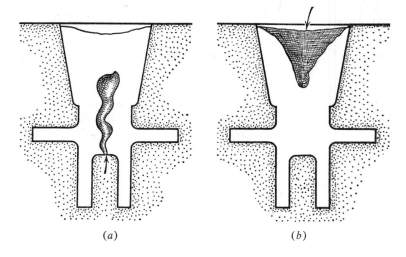

FIG. 6.5. Effect of atmospheric pressure on effectiveness of open risers: *left*, riser frozen over prematurely; *right*, riser kept open with insulating riser compound.

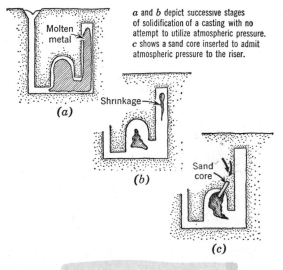

a and *b* depict successive stages of solidification of a casting with no attempt to utilize atmospheric pressure. *c* shows a sand core inserted to admit atmospheric pressure to the riser.

Molten metal

Shrinkage

Sand core

(a)

(b)

(c)

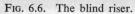

FIG. 6.6. The blind riser.

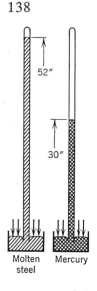

52"

30"

Molten   Mercury
steel

FIG. 6.7. The effect of atmospheric pressure on molten steel compared schematically with its effect on mercury.

could be delayed at the top of the riser, permitting atmospheric pressure to act on the liquid metal at this point.

Figure 6.6 illustrates how the atmospheric pressure principle may be used to feed castings "uphill." Figure 6.6*a* illustrates the initial skin formation of the casting and freezing off of the sprue. Figure 6.6*b* shows that, if the casting skin is not punctured, two central shrinkage cavities result, one in the riser, and the other in the casting proper. If, however, a "hot spot" is intentionally created in the riser, as by the small sand core pictured in Fig. 6.6*c*, atmospheric pressure acts through the sand core and "pushes" liquid metal from the riser up into the casting. Such a core is known as a *pencil core,* or *Williams core* and it is most commonly used on *blind risers* (risers totally enclosed in sand).

Blind risers may be used to feed steel castings to heights considerably higher than the top of the riser. The schematic barometer of Fig. 6.7 shows that air pressure should force liquid steel to a height of 52 in. in a perfect vacuum. In a solidifying casting an increasing partial vacuum tends to form as freezing and shrinkage take place, and the casting acts much as the barometer, with liquid metal being forced upward to prevent formation of a void. But the walls of the solidifying casting must develop enough strength to support the weight of metal to the height of the developing shrinkage. In steel castings the *skin strength* develops relatively rapidly, and feeding to superior heights is possible; this is *not possible for non-ferrous castings,* owing to slower development of adequate skin strength. In non-ferrous castings, the only way to prevent atmospheric dishing or puncturing at casting hot spots is to use risers (blind or open) which are higher than the casting itself; feeding must be *downhill.*

PART II. METHODS USED IN PRACTICE FOR CONTROLLING SOLIDIFICATION

6.6 Insulation

Insulating pads and sleeves for risers may be made of material which is insulating with respect to the material of the mold proper and which will withstand the corrosive action and temperature of the metal being cast. Plaster of paris, treated as described for plaster casting, is an excellent insulator for non-ferrous castings made in sand molds; sometimes sheet asbestos is used for insulating risers for magnesium castings. Plaster of paris cannot be used with ferrous castings, as the sulfur of the calcium sulfate reacts readily with iron at the high temperatures involved. For ferrous casting, perlite sleeves or fireclay-sawdust sleeves are sometimes used. In the latter case, the raw sleeve is fired before use to burn out the sawdust and leave air cells. Loose cellulose materials such as rice hulls are used for top insulation. These materials, however, are not nearly as effective an insulator as plaster of paris. Figure 6.8 shows typical applications of insulating pads and sleeves.

6.7 Moldable exothermic sleeves

Sometimes pads and riser sleeves are made from a moldable exothermic material which insulates by evolving heat spontaneously when ignited by the hot metal poured into the mold. The sleeves and pads are molded and used in the same manner as the insulating materials of Fig. 6.8.

6.8 Exothermic riser compounds

Exothermic riser compounds are loose materials sprinkled on the top of risers in place of insulating powder. They may be used not only to prevent heat loss from the top of the riser, but also to *add* some heat, and even to add some *molten metal* to the

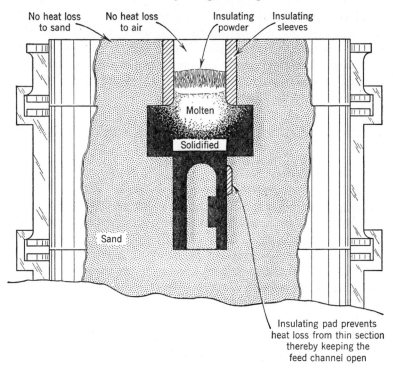

No heat loss to sand   No heat loss to air   Insulating powder   Insulating sleeves

Molten

Solidified

Sand

Insulating pad prevents heat loss from thin section thereby keeping the feed channel open

Fig. 6.8. Sketch showing how insulating pads and riser sleeves control direction of solidification.

riser. Exothermic riser compounds are essentially mixtures of a metal oxide and aluminum. When placed on the metal in an open riser, the mixtures react to give molten metal, heat, and a slag insulator. With iron oxide and aluminum, the reaction is

$$4Fe_2O_3 + 8Al = 4Al_2O_3 + 8Fe + 4500°F (2480°C) \text{ heat}$$

Oxides of other metals, such as copper, nickel, cobalt, and manganese, would react similarly. The molten metal produced by the reaction is an effective transfer medium for the very high heat, the sand around the riser is preheated, and the slag layer insulates the riser against heat loss to the atmosphere (Fig. 6.9). Accordingly, feeding efficiency is improved about 70 per cent, and risers can be made about a third the size of a normal riser. In fact, it is necessary to use short risers in order to concentrate the heat at the casting-riser neck where it is needed to prolong feeding and develop desirable thermal gradients in the casting. Additions can be made to the aluminum-metal oxide mixture to give metal of composition to match that of the castings; it is not really necessary to match the metal produced by the reaction to the composition of the casting as there is no tendency for the metals to mix. The density of the very hot metal produced exothermally is enough less (because of its high temperature) than that of the casting to discourage serious mixing.

6.9 Internal chills and chaplets

Internal chills are of several types (Fig. 6.10). They are used as in Fig. 6.11 to hasten solidification of a region of the casting inaccessible to a riser. *Internal chills* are usually of the same composition as the metal of the casting, and some surface

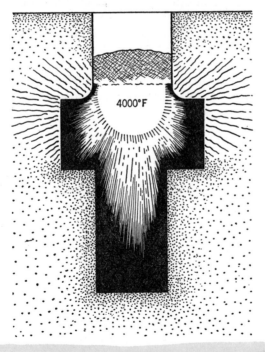

4000°F

FIG. 6.9. Action of an exothermic riser compound.

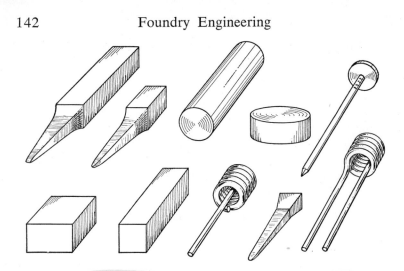

FIG. 6.10.   Typical internal and external chills.

melting is desirable in order to fuse the chill into the body of the casting over its entire surface; this involves careful selection of size and shape, as the chill must not melt prematurely and become dislocated.   Chaplets, used to support cores as in Fig. 6.12, can be considered as internal chills for the purposes of this discussion; the main difference is that in chaplets surface fusion is desirable, but the chaplet must remain strong enough to support

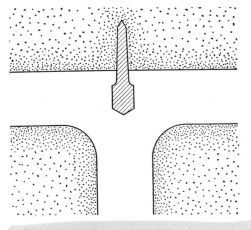

FIG. 6.11.   Typical use of an internal chill.

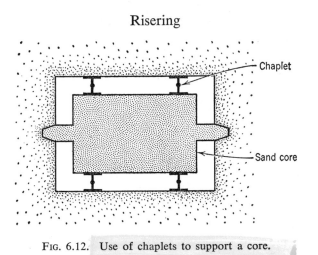

FIG. 6.12. Use of chaplets to support a core.

the core.   Internal chills only need fuse properly and support
their own weight.   Usually, foundrymen do not use internal chills
and chaplets unless absolutely necessary because *blows* are often
found in regions where they are used.   Actually, when properly
engineered, internal chills and chaplets are beneficial and safe.
It is necessary to select the proper size and shape, and to keep
the chill or chaplet clean and dry.   Internal chills and chaplets
are usually coated with tin or some other material by the supplier
to prevent rusting.   The coating is no guarantee of cleanliness.
The ideal chill would be one heated overnight at 800°F (425°C)
to remove gases, sandblasted before use, and kept scrupulously
dry and clean to prevent blowholes in the casting.   If the chill
is to remain in a green sand mold more than an hour or so, it
would have to be wrapped in Pliofilm or aluminum foil to prevent condensation or corrosion on the chill or chaplet.   Internal
chills or chaplets with fancy stems do not provide a particularly
good surface for fusion; the metal often does not fill complicated
contours in which it is easy for gas and moisture to collect.
Internal chills and chaplets are seldom used for non-ferrous castings because of the difficulty of obtaining proper fusion in these
metals.

6.10   External chills

External chills need not be of the same material as that of the
casting since fusion must not occur.   Iron and steel chills are used

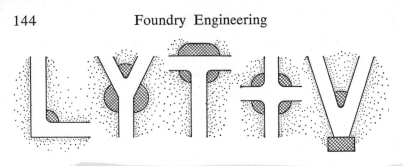

Fig. 6.13. Use of external chills to prevent the formation of hot spots at casting junctions.

for ferrous castings, and these and copper and block graphite chills are used for non-ferrous castings.

Valuable results are obtained by using external chills to hasten solidification; foundrymen do not utilize this expedient to full advantage. External chills are excellent for controlling cooling rates in critical regions of castings. Figure 6.13 shows a few typical applications. When properly used, external chills are fully dependable and safe. To obtain the best performance from external chills, it is advisable to have a chill engineer whose job it is to determine chill shapes, sizes, and general handling and to have proper chills for particular jobs. Many otherwise difficult or expensive feeding jobs can be alleviated by good chill technique. Like internal chills, external chills should be clean and dry; sandblasting just before use is good practice if metal chills are used. Highly polished chills are not good, as the metal does not lie well to the smooth surface; a few scratches on a polished chill help, but sandblasting is best. A layer of Cellophane over the chill surface will protect it against rusting or condensation of moisture when left in a green sand mold overnight. Block graphite is an excellent chill material. The conductivity of graphite is about three times that of steel, but specific heat is only one-third; this means that a graphite chill should be at least three times as large as a steel chill to do the same job. External metal chills can be used on irregular contoured shapes as well as on flat surfaces; it is only necessary to cast a metal chill to the shape desired.

6.11 Molding materials of different chill capacities

A mold made as a composite of different molding materials offers excellent opportunity for controlling solidification. For example, zircon sand is often used on internal angles to hasten freezing at the hot spot; ground chrome and chrome magnesite are other possibilities for mild chilling. Figure 6.14 is a schematic interpretation of the possibilities.

6.12 Padding

Figure 6.15 shows typical examples of padding for improving casting soundness. Because the casting is thinnest at points distant from the riser, freezing is complete in these regions first. The sections are made gradually thicker toward the riser so that freezing progresses directionally, and each successive region to freeze is supplied with enough molten metal to feed it properly. This *padding* can be removed by machining, or the casting can be designed with the tapered sections as integral, useful parts of the ultimate assembly. Usually padding is an extreme expedient, as engineers do not like the added weight; in the long run, however, much would be saved by making tapered designs the rule (especially for steel castings) rather than the exception. Padding should be provided only after due consultation to determine how the casting is to be made; otherwise the taper may be placed in the wrong direction.

6.13 Riser shape, size, and contact area

Round risers are considered to be more efficient than square or rectangular risers; the ratio of surface area to volume is lower, and heat loss to the molding material would seemingly be less.

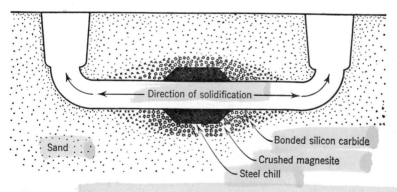

Direction of solidification

Bonded silicon carbide

Crushed magnesite

Steel chill

Sand

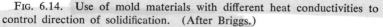

FIG. 6.14. Use of mold materials with different heat conductivities to control direction of solidification. (After Briggs.)

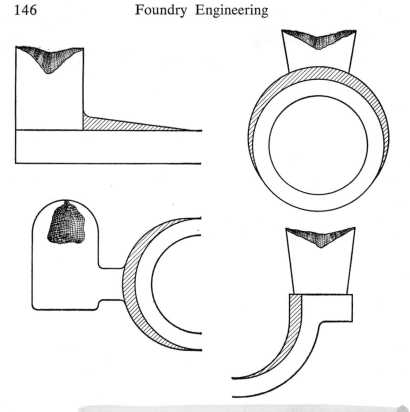

FIG. 6.15. Typical examples of padding castings to obtain directional solidification.

Accordingly, the effective feeding life of round risers should be best, and foundrymen usually use round risers whenever possible. Spherical risers would have the lowest ratio of surface area to volume, but they would be impractical. Blind risers with a hemispherical top and cylindrical body approach what is considered optimum shape.

Risers must be at least 15 to 20 per cent larger in diameter than the section they are to feed (unless extensive chilling or insulation are employed). The riser height never needs to be greater than 1.5 times the diameter and can usually be less. Short, fat, top risers are more efficient than tall, slim ones of the same volume.

The design of neck and contact area is particularly important. Figure 6.16 shows examples of good and bad designs. The

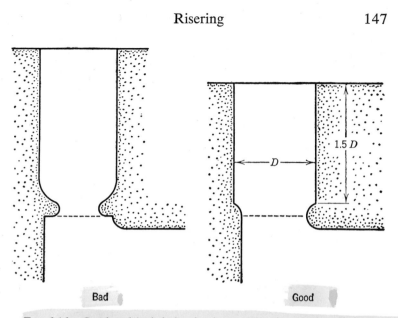

Bad   Good

FIG. 6.16. Good and bad design in risers. The riser at the left is proportionately too tall and has inadequate contact area.

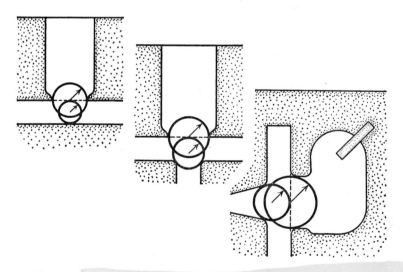

FIG. 6.17. Determining directional solidification by method of inscribed circles.

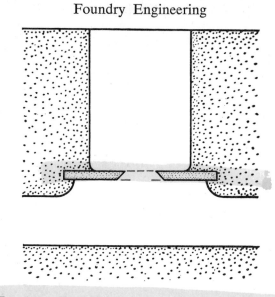

FIG. 6.18. "Necked-down" riser with "knock-off" core.

contact area can be determined by the method of inscribed circles shown in Fig. 6.17. Knock-off (or neck-down) cores (Fig. 6.18) may be used under risers to reduce the cost of removing them from the casting; the core is preheated by metal flowing over and through it and is so thin that very little heat is conducted into the mold. The necked-down contact area remains hot and allows molten metal to feed through until the casting is frozen. Then the riser can be flogged or cut from the casting.

6.14 Side and blind risers

It is frequently advisable to use *side risers* instead of *top risers*. The side risers may either be *blind* (totally enclosed in sand) or *open* (riser extending up to cope surface). Whether blind or open, side risers for cast iron and non-ferrous castings must extend higher than the castings as the skin strength of these materials does not permit feeding "uphill." A blind riser for feeding steel "uphill" is illustrated in Fig. 6.6.

Blind risers have certain advantages over top, open risers:

1. Bottom gating through the side riser provides proper temperature gradients; the last metal to enter, and the hottest, is at

the bottom of the casting and in the riser which provides conditions for good feeding; also the sand in these regions is preheated.

2. Blind risers can be placed at any position in the mold.

3. The cylindrical riser with a hemispherical dome approaches a minimum ratio of surface area to volume and thus slowest cooling.

4. Blind risers can be connected to the flat side of a casting, as in Fig. 6.19, and can be removed more easily than if connected to a contoured surface, as top, open risers usually are.

6.15 Location of risers

Enough has already been learned to indicate that the location of risers on a solidifying casting is an important (and often difficult) consideration. In Section 6.4 it was shown that risers must be placed so that solidification is *directional* from the casting extremities toward the riser.

Efficiency of feeding, however, must be balanced against cost of removing risers from the casting. It would not be difficult to locate all enlarged areas of a casting and place a riser at each position. To do so, however, might result in as little as 30 per cent metal in the casting proper, 70 per cent in risers, and a high

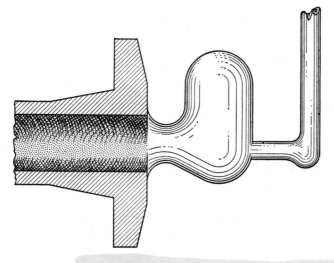

FIG. 6.19.   Blind riser attached to a padded flanged fitting.

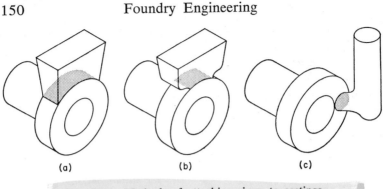

(a)   (b)   (c)

FIG. 6.20. Methods of attaching risers to castings.

cleaning cost. (This suggests care in designing the original casting so that it can be *fed* with a minimum number of risers.) The manner of attaching the riser to the casting affects the cost of its removal. Figure 6.20 shows three examples of how risers may be attached; it is obvious that more work is required to remove the riser of example *a*, less for example *b* and still less for *c* (since it is on a flat surface). In locating risers it should be remembered that:

1. Directional solidification must be maintained. This means that heavy sections cannot be fed through thin sections without the use of insulating or heating pads, or chills. It also means that there is a definite limit to the distance a riser can feed even in a uniform section. (This distance is often called *feeding distance*.)

2. Risers attached to flat, open sections are easier to remove than those attached to curved sections.

3. Gating through risers not only improves thermal gradients but also eliminates the need for cutting the gate and cleaning the gate entrance area.

4. Location of risers influences quality and cost of castings, and so is a serious consideration. Each casting should be carefully analyzed for risering by a competent engineer, and no compromise made with requirements.

6.16   General considerations of risering

For greatest efficiency, risers should be round rather than square; cooling is minimized, owing to a lower ratio of surface area to volume.

All risers function best when kept open to the atmosphere; this insures that the riser will not draw metal from the casting as the result of a partial vacuum in the riser.

The feeding efficiency of top, open risers improves for a given height of casting as section size increases; in other words, the amount of taper or padding needed to make a sound casting decreases with increasing section size. One reason for this is that convection currents tend to form when hot metal is at the bottom of a casting, as in bottom gating, and cold metal is at the top. In thin castings, freezing is too rapid and the metal becomes sluggish too quickly to give time for the establishment of effective currents. Convection currents in molten metal are not unlike those in air in a room, where the warm air rises to replace colder air at the ceiling.

Feeding can be prolonged by pouring molten metal into risers at suitable intervals after the mold is first filled, a practice called "visiting" risers. The addition of the hot metal delays freezing, but visiting is effective only with large castings.

Side risers are desirable, particularly if the casting is gated through them. The temperature gradients are established favorably by natural methods. A blind riser is efficient because it represents the minimum practical ratio of surface area to volume and so has a slow cooling rate. In steels, blind risers can be used for feeding sections of castings higher than the riser itself, because a strong solid skin rapidly freezes which can support the gravity differential. In cast iron and non-ferrous castings, side risers must be depended upon only to feed regions of the casting below the effective liquid metal head in the riser; such metals do not develop skin strength fast enough to support feeding to superior heights. It is always desirable to use sand or graphite cores to keep blind risers open to the atmosphere. Figure 6.19 shows a typical use of blind side risers.

PART III. THEORETICAL CONSIDERATIONS OF RISERING

6.17 Riser size

Discussion in the previous portions of this chapter has been limited to general principles for risering castings. Some of these principles have been well known to foundrymen for many years;

others are comparatively new; all of them comprise the bases for risering commercial castings. In general, the principles that have been described are useful only in a qualitative sense, and require practical experience for their proper application to obtain sound castings with a minimum of wasted metal (highest yield). In very recent years, however, a firm foundation has been laid in theoretical analyses of the solidification and feeding processes, and in many instances a more quantitative engineering approach to the risering of commercial castings has been possible.

The earliest and most widely known quantitative risering analysis is that of Chvorinov.[1] Chvorinov showed that the time for complete solidification of a cast shape is proportional to the square of the volume of the casting divided by the square of the surface area of the casting, or:

$$\theta_f = K \left(\frac{V}{A}\right)^2$$

(1)

where θ_f = freezing time
V = volume of casting
A = surface area of casting
K = constant

A theoretical justification of this equation has already been indicated (Chapter 5); it was pointed out there that the equation is not absolutely rigorous for all castings, but is a good engineering approximation for simple shapes.

One of the requirements of an adequate riser is that it must remain molten longer than the casting it is to feed. With this in mind, Chvorinov's rule serves as a rough guide in determining riser sizes. Castings with a large volume-to-area ratio (spheres or cubes), freeze relatively slowly and so require large risers in comparison to their size. Castings of the *same volume,* but with a lower volume-to-area ratio (plates, bars) freeze more quickly, and so a smaller riser may be used to obtain adequate feeding. For example, in casting a 100-lb steel cube, a riser of about 100 lb is required (unless special riser treatments are employed), but to feed a 1-in.-thick steel plate weighing 100 lb, a riser weighing only 15 lb is of adequate size.

Attempts have been made to calculate riser sizes (using

Chvorinov's rule) by considering the riser and casting as two separate castings, and determining a riser size such that

$$\left(\frac{V}{A}\right)_{\text{Riser}} > \left(\frac{V}{A}\right)_{\text{Casting}}$$

Such attempts have not been successful, however, because Chvorinov's rule takes no account of solidification shrinkage. As an example, consider the three shapes shown in Fig. 6.21a. The first shape is a cylinder 4 in. in diameter by 4 in. high, insulated so that heat can pass only into the circumferential sand walls. With a K value of 13.7 minutes per square foot (approximately correct for steel in green sand), Chvorinov's rule may be used to calculate the freezing time for the cylinder (13.7 minutes). The freezing time of a 4-in. steel cube in the same sand is 6.1 minutes, and the freezing time of a 2-in. plate, 8 in. square is also 6.1 minutes. Since the calculated solidification time of the cylinder is nearly twice as long as that of either the plate or the cube, it might be expected that if the cylinder were attached to these castings it would serve as an adequate riser. The result is shown in Fig. 6.21b. The riser does, in fact, feed the cube, but the plate contains a large shrinkage cavity. The cavity develops because the plate, being much larger than the cube of equivalent freezing time, required much more feed metal to feed *solidification shrinkage*. In effect, the casting drained the riser dry early in the freezing process. We may now reiterate the two requirements governing riser size, which were mentioned early in this chapter:

1. The riser must be large enough to freeze after the casting; that is, riser size and shape must be adjusted so that its volume-to-area ratio is at least greater than that of the casting.

2. The riser must possess enough liquid to feed the casting. Liquid metal disappears from the riser in one of two ways. It freezes to form solid metal in the riser; hence item 1 above is important. It also is drained from the riser by the casting solidification shrinkage, and so the riser must be large enough to possess extra liquid metal for this purpose.

The first workable application of these two principles to actual calculation of riser sizes was that of Caine.[2,3] Starting from the above propositions Caine developed his risering curve (for steel castings) shown in Fig. 6.22. The curve is in part theoretical since the parameters used for the plot were clearly suggested by

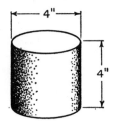

θ_f = 13.7 minutes

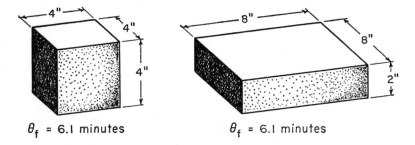

θ_f = 6.1 minutes θ_f = 6.1 minutes

FIG. 6.21a.   Solidification times of some simple castings (steel in green sand). Cylinder is insulated at ends, θ_f = solidification time.

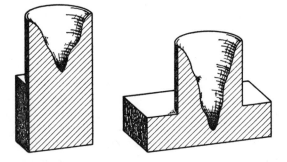

FIG. 6.21b.   Risering a cube and plate. Both castings have equal freezing times; the riser which is adequate to feed the cube is not large enough to feed the plate. Riser is 4 in. in diameter by 4 in. high insulated on top; cube is 4 by 4 in.; plate is 8 by 8 by 2 in.

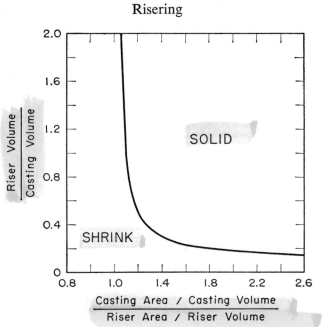

FIG. 6.22.   Relative riser and casting geometry to obtain sound steel castings.   (After Caine.[9])

Chvòrinov's rule, and the two extreme end points of the curve were suggested by the second requirement above, that a riser possess enough liquid metal to feed the casting shrinkage.   The actual shape of the curve, however, is empirical, having been determined from experimental data.

The great value of Caine's curve lies in the fact that it permits plotting data from a wide variety of casting and riser shapes and sizes on a single set of co-ordinates.   Conversely, it permits judicious choice of riser size for a wide variety of casting shapes and sizes.   The curve shows that as the volume to surface area of a casting increases (with respect to that of the riser) the required riser volume decreases.   The curve shows further that for a casting of a given volume there is a minimum riser size required (at the far right of the curve) which is that amount of metal required to feed solidification shrinkage.

Caine's curves have been used for many years in steel foundry practice.   Often they are used on rather complex shapes where

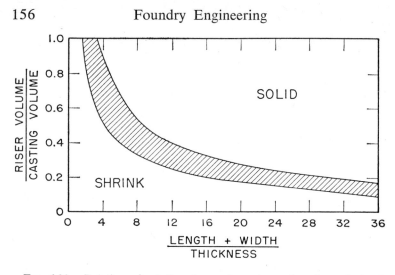

FIG. 6.23. Relation of relative riser volume to casting shape factor to obtain sound steel castings. For risers with height to diameter ratios of 0.5 to 1. (From Myskowski, et al.[4])

more than one riser is required. In these cases, the size of each riser is calculated separately on the basis of the volume and surface area of the *portion of the casting the riser is intended to feed.* Recently, modifications of Caine's curves have been developed to permit simpler and more rapid calculation of riser sizes, and to allow for section changes, hot spots, etc.[4] One such curve is shown in Fig. 6.23 for steel castings.

Most applications of curves of the above type have been in the field of steel castings. There appears to be no reason, however, why the general approach should not be valid for other metals as well. Figure 6.24 illustrates application of the method to risering hypoeutectic ductile iron.

A more basic approach to the problem of risering has been undertaken by Adams and Taylor.[5] In this study an entirely theoretical approach was used to predict riser behavior from fundamental thermal and solidification constants. The risering equation developed for the simplest case of a blind riser (no chills or insulating material) is as follows:

$$(1 - \beta) \frac{V_R}{V_C} = \frac{A_R}{A_C} + \beta \qquad (2)$$

where β = solidification shrinkage (about 0.03 for steel)
     V_R = riser volume
     V_C = casting volume
     A_R = riser surface area
     A_C = casting surface area

To date only a limited amount of experimental work has been conducted to verify this and related equations. Comparison with the work of Caine, however, has indicated close agreement between

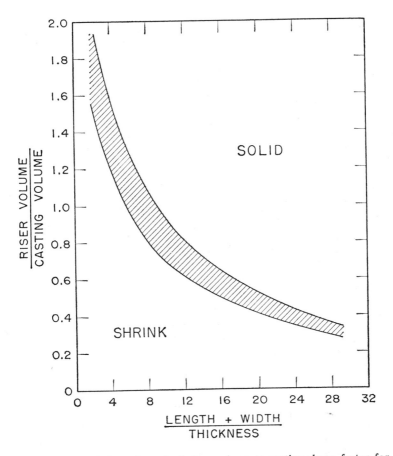

FIG. 6.24. Relation of required riser volume to casting shape factor for a hypoeutectic ductile iron (2.65–3.0 per cent C, 2.46–2.78 per cent Si, 3.55–3.82 per cent carbon equivalent). (From Bishop and Ackerlind.[8])

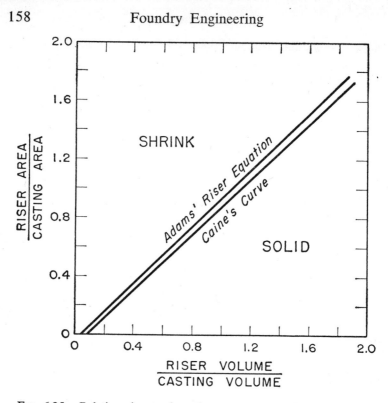

FIG. 6.25. Relative riser and casting geometry to obtain sound steel castings. Lower curve replotted from Caine's data for top blind risered castings [2]; upper curve calculated from Adams' riser equation. (From Adams and Taylor.[5])

theory and practice (for steel castings). Figure 6.25 plots Caine's data (for top, blind risered castings) and compares with it the curve obtained from equation 2. The co-ordinates used are those of Adams.[5]

The essential advantages of the more theoretical approach described above are that (1) it should be applicable to a variety of metals, not steel alone, and (2) with minor modifications, it can be adapted for prediction of the effects of chills, riser insulation, radiation losses from top risers, etc., on the behavior of a riser. For a full treatment of these variables, the reader is referred to the reference cited above.

6.18 Riser treatments

A variety of methods of improving the efficiency of a riser (increasing casting yield) have been cited in earlier portions of this chapter. The methods included *insulating* the walls of the riser, insulating the top of the riser (*shielding*), use of exothermic materials, etc. It is our purpose here to describe the results of some theoretical analyses of heat flow in risers, and cite some practical conclusions of these studies regarding the insulating and shielding of risers.

Heat is lost from an open riser in three ways: by convection and by radiation from the top, and by conduction into the sand walls surrounding the riser. Of these, radiation and conduction are by far the most important, but there is no simple relationship between the amount of heat lost by each mechanism; the relative amounts depend on a number of factors, including metal solidification temperature, riser size, and riser shape. For example, Fig. 6.26 illustrates the effect of riser size on heat losses from the top of a steel riser. The larger the riser, the greater the percentage of heat that flows out the top. Figure 6.27 illustrates how the

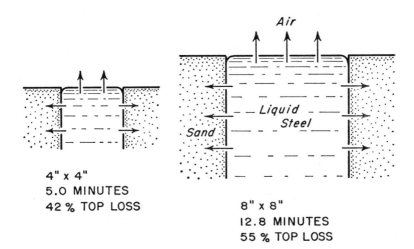

4" x 4"
5.0 MINUTES
42 % TOP LOSS

8" x 8"
12.8 MINUTES
55 % TOP LOSS

Fig. 6.26. Size effect on heat loss from tops of open risers. Top losses are expressed as a percentage of total heat lost from the risers during solidification. Risers are cylindrical. (From Adams and Taylor.[6])

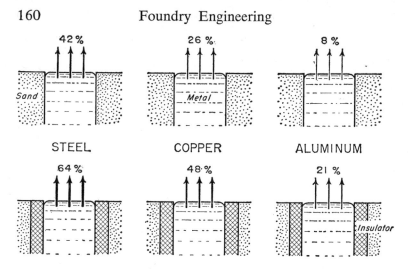

FIG. 6.27. Top heat losses from risers of several metals. Risers are 4 in. in diameter by 4 in. high; insulator has thermal properties similar to gypsum. (From Adams and Taylor.[6])

percentage of top losses varies with metal poured. The higher-melting metals lose a greater proportion of their heat by radiation. In the 4×4-in. riser, surrounded by sand and unshielded, steel loses 42 per cent of its heat by radiation, while copper and aluminum lose only 26 per cent and 8 per cent, respectively. If an insulating sleeve whose thermal properties are equivalent to those of gypsum is placed around each of the risers, but the risers are left unshielded, a greater percentage of heat is lost from the top. Steel, copper, and aluminum lose 64 per cent, 48 per cent, and 21 per cent, respectively.[6]

The practical implications of such riser heat-flow analyses are evident in Fig. 6.28. Unshielded risers of steel, copper, and aluminum freeze in 5.0, 8.2, and 12.3 minutes, respectively. Under no condition can these risers feed a casting that requires longer than these times to freeze. If, however, the riser heat losses can be reduced, it becomes possible to feed larger sections.

Since open steel risers lose such a large portion of their heat by radiation to the atmosphere, top insulation (*shielding*) is very effective in reducing over-all heat losses, and thereby increasing riser solidification time. Figure 6.28b indicates that shielding a 4-in.-diameter by 4-in.-high riser increases its solidification time by about a factor of 3. The increase would be even greater for the 8-in.-diameter by 8-in.-high riser. Shielding is less effective on the copper riser, and much less effective on the aluminum riser, since these metals lose a relatively smaller amount of heat out the top. Insulating the side walls of aluminum and copper risers (Fig. 6.28*c* results in substantial improvement in riser solidification time as would be expected from the foregoing. From a heat-flow standpoint the optimum condition for all metals is, of course, a combination of insulation and shielding (Fig. 6.28*d*).

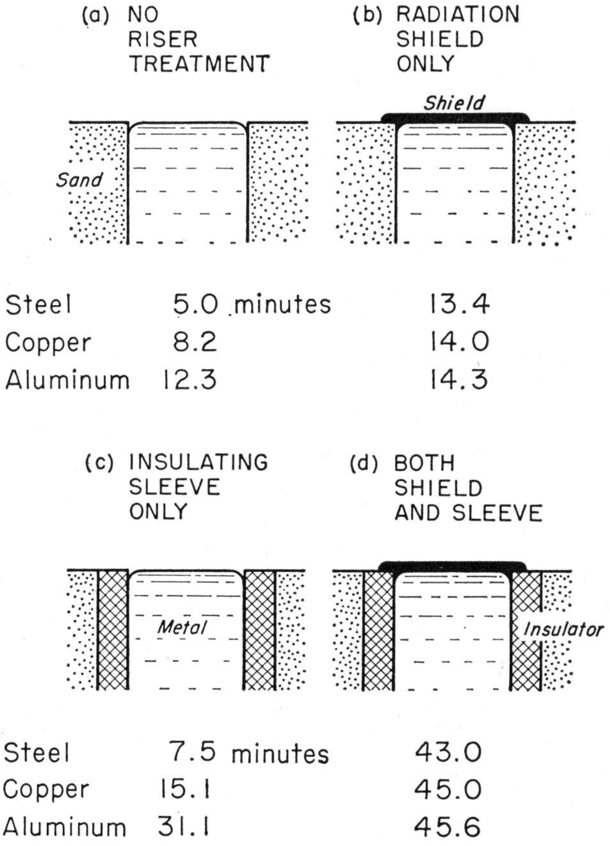
(a) NO RISER TREATMENT
(b) RADIATION SHIELD ONLY
Shield
Sand
Steel 5.0 minutes 13.4
Copper 8.2 14.0
Aluminum 12.3 14.3
(c) INSULATING SLEEVE ONLY
(d) BOTH SHIELD AND SLEEVE
Metal
Insulator
Steel 7.5 minutes 43.0
Copper 15.1 45.0
Aluminum 31.1 45.6

FIG. 6.28. Time in minutes for total solidification of 4-by-4-in. cylindrical risers. Insulator has thermal properties similar to gypsum; perfect radiation shield assumed for (*b*) and (*d*). (From Adams and Taylor.[6])

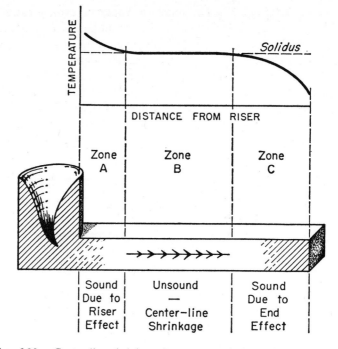

Fig. 6.29. Center-line shrinkage in a cast steel plate. Top, temperature distribution in plate near end of solidification; bottom, center-line shrinkage in area of the temperature plateau.

6.19 Riser feeding distance

In order to properly feed a casting, a riser must not only be of a certain minimum size; it must also be in the proper *location*. Stated differently, the riser must maintain a positive pressure of liquid metal to all portions of the casting it is to feed. Figure 6.4 showed an example where this pressure was not maintained. The riser was adequate in size but was not placed so that the liquid feed metal reached all portions of the casting during solidification; *center-line shrinkage resulted.*

Figure 6.29 illustrates how center-line shrinkage develops in a plate casting (0.30 per cent carbon steel). The temperatures at various portions of the casting (near the end of solidification) are shown at the top of the figure. Note that the casting consists of three zones, *A*, *B*, and *C*. In zone *A*, a thermal gradient exists owing to the heat effect of the riser. In zone C a similar gradient exists owing to the casting end effect. In these zones, solidification proceeds *directionally* toward the riser, and in the final casting the zones are free of shrinkage defects. In the central zone (zone B) a temperature "plateau" exists. Freezing in this zone is not directional; the dendrites interlock near the end of solidification, cutting off feeding, and center-line shrinkage results. It has been found that, if the thermal gradients at any point along the length of a plate are less than about 1 to 2°F per in., center-line shrinkage results. By reducing the length of the plate until the

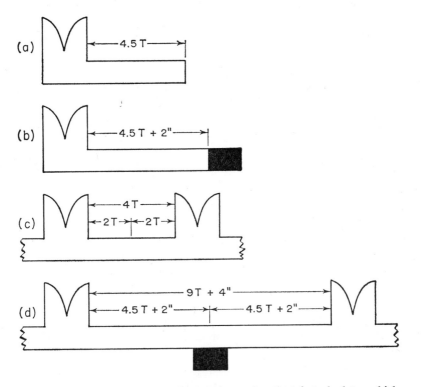

FIG. 6.30. Sketches illustrating maximum lengths of steel plates which can be cast free of center-line shrinkage. (*a*) Sand-cast plate. (*b*) End-chilled sand-cast plate. (*c*) Double-risered plate. (*d*) Double-risered plate chilled at center. T represents plate thickness in inches. Steel analyses: 0.25/0.30 per cent C, 0.05/0.10 per cent Mn, 0.35/0.50 per cent Si. (Data from Pellini.[10])

combination of the riser heating effect and the end effect produce a thermal gradient over the entire length, a sound plate may be produced.

Extensive research on this general subject has been conducted over the past 5 years at the Naval Research Laboratory. Much useful design data have been obtained on the ability of risers (in steel castings) to feed various casting section sizes. Figure 6.30 illustrates an example of some of the data which applies to simple plate sections. The data show that there is a maximum length plate which can be fed with a simple riser (4½ times the section thickness), but that this distance can be increased by end-chilling the casting, or by using more than one riser in combination with chilling. An alternative method would be to *pad* the plate as described earlier. The bulk of the Naval Research Laboratory data have recently been summarized and references to earlier work are contained therein.[7, 10]

6.20 Risering of alloys

Discussion of risering theory has heretofore been limited primarily to pure metals, or to nearly pure metals such as steel. The principles described apply also to commercial foundry alloys such as brasses and bronzes, and aluminum and magnesium alloys. However, in risering these alloys it is well to remember their completely different solidification characteristics, described in Chapter 5. With a few exceptions, the alloys freeze with such a wide "mushy" zone that they are not prone to the center-line shrinkage defects commonly found in improperly risered pure metal or steel castings. Rather, they are subject to *microporosity*. In improperly risered castings the microporosity can be so concentrated that it is readily visible on machined surfaces; "leaky" castings with low mechanical properties result.

The risering requirements of alloys vary widely from one to another. This undoubtedly depends somewhat on the presence or absence of minute, invisible micropores in the cast metal. In part it is also due to the fact that some alloys have a smaller liquid-solid shrinkage than others. A systematic study of shrinkage of alloys has, unfortunately, not yet been conducted. Practical experience is still a prerequisite in their efficient risering, but it should be re-emphasized that this efficient risering is possible only

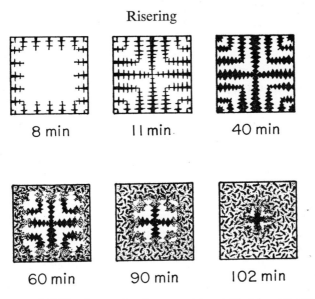

8 min   11 min   40 min

60 min   90 min   102 min

Fig. 6.31. Solidification mode of gray cast iron. Sketches are cross sections of a 7 in. square bar at various times during solidification. (After Bishop and Pellini.[11])

if a basic understanding is obtained of the risering principles heretofore described.

From a solidification and risering standpoint, perhaps the most interesting and certainly the most unusual alloy is gray cast iron. Figure 6.31 illustrates schematically the solidification of hypoeutectic gray iron (cast in a bar of 7 in. cross section). Early in the solidification process, primary austenite dendrites form and then reach to the center of the casting after 11 minutes. The casting is "mushy" throughout, but adequate feed channels are still available to compensate for solidification shrinkage. After nearly 60 minutes, the entire cross section is at the eutectic temperature; the eutectie (flake graphite plus austenite) begins to form at the casting surface and grow inward. At 102 minutes after pouring, the eutectic solidification has progressed nearly to the center of the bar; the casting is almost completely solid.

During solidification of the eutectic, demand for feed metal essentially ceases and, depending upon chemical composition and mold variables, the liquid metal may even flow from the casting back to the riser. This surprising effect occurs because when graphite precipitates from molten iron it actually expands, and, if enough carbon is free to precipitate, the over-all volume may *increase* during solidification rather than decrease. Because the graphite expansion in cast iron can in large measure counterbalance solidification shrinkage, it is possible to feed this alloy with much smaller risers than are needed for other metals. Under proper conditions gray iron may require little risering or no risering at all; solidification shrinkage (except in very heavy section castings) can be fed with small riser "bobs" or through the gate itself.

In spite of the foregoing, risering of gray iron castings is not always simple. Paramount among the problems is *dilation,* the "swelling" of a casting during solidification to a size greater than that of the original pattern. Besides resulting in oversize castings, dilation can result in so much added demand for liquid feed metal (during freezing) that ordinary risers are wholly inadequate. Figure 6.32 illustrates the effects of dilation on riser requirements of a cast iron cylinder. In dry sand, dilation does not occur; the graphite expansion counterbalances solidification contraction, and the cast iron "shrinks" very little (Fig. 6.32a). In green sand, however, when the casting is kept open to the atmosphere, dilation results in a cylinder which is bulged slightly outward; gross internal shrinkage is the result (Fig. 6.32b). When atmospheric pressure cannot act through an insulated "hot spot," the outward dilation is balanced by inward contraction (Fig. 6.32c) and no gross shrinkage results. Figure 6.32d illustrates that very large risers may not be sufficient to feed shrinkage in a green sand casting and may, in fact, accentuate shrinkage by increasing the ferrostatic head and the "hot spot." On the other hand, very small risers completely feed a dry sand casting (Fig. 6.32d).

In practice, the factors that affect casting dilation, and hence the amount of gross shrinkage that occurs are (1) the size and shape of the casting, (2) the action of atmospheric or ferrostatic pressure, (3) the characteristics of the molding material, (4) the size, shape, location, and thermal treatment of gates and risers, and (5) metal chemistry. The fundamental cause of the dilation itself is thought to be the graphite expansion "pushing" against the mold wall.[13] If true, the surprising situation is encountered that graphite expansion can either *reduce* shrinkage by expanding inward; or, when improperly controlled, can *accentuate* shrinkage

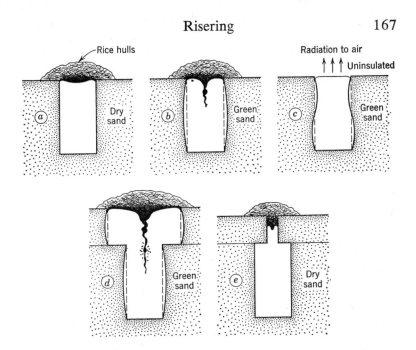

Fig. 6.32. Schematic cross sections of a gray iron cylinder solidified under various molding and risering conditions. Cylinders are 8 in. high by 3¾ in. in diameter. Dilation is exaggerated for illustrative purposes; it is usually less than about ⅟₃₂ in. in castings of this size.

by buckling outward. There is much yet to be learned about the solidification and risering of this, one of the foundry's oldest cast metals.

6.21   Summary

This chapter was divided into three parts; the first was a general discussion of risering, the second was a description of risering and chilling methods used in practice, and the third was devoted to theoretical analyses of risering.

In the first section it was pointed out that nearly all cast metals must be *risered* to avoid shrinkage defects. An adequate riser for this purpose is one that is large enough, and is placed in such a location that solidification in the casting proceeds *directionally* toward a pool of readily available liquid metal (in the riser). Also, the riser should be so designed and treated that it accomplishes its task in the most economical manner possible.

A variety of methods are used in practice to assure that directional solidification toward a riser is obtained, and to permit reduction of the size of the riser required. These include use of (1) insulating materials, (2) exothermic materials, (3) internal and external chills, (4) moldable chills, (5) padding, (6) careful design of riser and riser contact area, and (7) careful design of riser location. These methods are discussed in the second part of the chapter, and Section 6.16 summarizes some general considerations for practical risering.

In very recent years a firm foundation has been laid in theoretical analyses of the solidification and feeding processes. In many instances these analyses permit a quantitative engineering approach to the risering of commercial castings, where previously experience was the only guide. Several analyses have been conducted on the riser size required to feed steel castings; practical experience has shown these analyses to be valid and useful. Thermal analyses of steel castings have revealed the factors affecting center-line shrinkage in steel castings and have resulted in design rules for locating risers on steel castings. Theoretical analyses of the effects of various riser treatments (insulation, etc.) have shown quantitatively the great benefits to be gained by these treatments. The third part of this chapter has been devoted to discussion of a variety of theoretical studies of risering, including those outlined above. These studies have been included because of the immediate practical value of many of the conclusions of the studies, and because they point the way to future methods of producing sounder, stronger castings by more economical methods.

6.22   Problems

1. What factors determine riser adequacy? What factors determine riser economy?

2. What factors would you consider in determining riser location on a casting?

3. Show by calculation the theoretical height to which a blind riser can feed a steel casting under ideal conditions.

4. What factors limit the use of blind risers?

5. What is the purpose of the sand or graphite core in blind risers?

6. Draw isotherms indicating progress of solidification in a satisfactory casting-riser combination. Repeat for an unsatisfactory combination.

7. What is the ideal theoretical shape of a riser? The best practical shape?

8. Discuss briefly factors that must be considered in determining the size and contact areas of risers.

9. Why do necked-down risers function thermally as well as ordinary risers in spite of the thin sand core at the bottom?

10. Discuss the mechanism of riser feeding. How do blind risers feed metal? What are four advantages of blind risers over conventional top risers?

11. Show with sketches four different effects of shrinkage. How can they be eliminated? Illustrate these methods simply.

12. List seven ways for developing proper temperature gradients; comment on each briefly.

13. Why are riser compounds used? Illustrate with sketches the effects of riser compounds.

14. Compare critically the solidification of risers treated with insulating and with exothermic materials.

15. What undesirable results can sometimes be obtained when using chills in green sand molds? Can these undesirable effects be avoided? Elaborate on your answer.

16. Discuss briefly why fireclay and diatomaceous earth should prove satisfactory as material from which to make insulating riser sleeves.

17. Why should moldable exothermic riser sleeves be so desirable in the foundry? Why should they prove even more valuable on light metals than on steel? Give a reasonably quantitative answer based upon thermal considerations.

18. What types of exothermic materials would you use for making nickel, iron, and steel castings? Would it be possible to use a single type exothermic material for two of these metals? Explain your answer.

19. Why are short, thick risers required when using exothermic compounds? What effect do convection currents have on riser operation? How would this effect vary for open-top risers and blind side risers?

20. Show by sketches how solidification can be controlled by padding. What effect does padding have upon bottom-gated, top-risered castings?

21. How would you increase efficiency of an open riser?

22. Discuss the formation of interdendritic shrinkage. Why is it so pronounced in certain types of alloys? What steps can be taken to eliminate interdendritic shrinkage?

23. Discuss reasons for so-called *leakers*. How may this difficulty be eliminated? In what metals is it most common? Obtain a scrapped

casting of an alloy known as "85–5–5–5" from a local foundry, and make a typical non-destructive X-ray of one segment; then cut a thin slab from this same section and make a transverse X-ray. Note the greater value of the transverse radiograph as a critical inspection tool.

24. Calculate the most efficient height-to-diameter ratio for a top blind cylindrical riser. Assume that heat flow per unit area is the same for the sides and top of the riser.

25. An aluminum sand casting is rigged with an open-top riser. Optimum yield is obtained with the untreated riser when the riser is 4 in. in diameter by 6 in. high. For maximum yield in each of the following cases would you increase or decrease the *height-to-diameter ratio?*

(*a*) Riser insulated with sleeve only.

(*b*) Riser insulated on top only.

(*c*) Entire casting and riser poured in a plaster mold (open-top riser).

(*d*) Casting poured of copper in a sand mold (open-top riser, untreated).

26. A 5-in. cube of pure copper is cast in green sand with a 5-in.-diameter cylindrical top riser. Both riser and casting are completely enclosed in sand. How high must the riser be to produce a sound casting? Assume that solidification shrinkage for copper is 4.9 per cent.

27. A 5-in. cube of 0.30 per cent carbon steel is cast in green sand with a 5-in.-diameter cylindrical top riser. Both casting and riser are completely enclosed in sand. How high must the riser be to produce a sound casting?

28. Repeat the calculation of Problem 27 for a steel plate the same weight as the 5-in. cube ($10 \times 12.5 \times 1$ in.). Would you recommend using the 5-in.-diameter riser on this casting?

29. Where would you place the riser on the casting of Problem 28?

30. Discuss the formation of center-line shrinkage. Why doesn't it occur in most aluminum alloys?

6.23 Reference reading

1. Chvorinov, N., "Control of the Solidification of Castings by Calculation," *Foundry Trade Journal*, 95–98, August 10, 1939.

2. Caine, J. B., *S.F.S.A. Research Report*, **13**, 1947.

3. Caine, J. B., "A Theoretical Approach to the Problem of Dimensioning Risers," *Transactions A.F.S.*, **56**, 492–501, 1948.

4. Myskowski, E. T., Bishop, H. F., and Pellini, W. S., "A Simplified Method of Determining Riser Dimensions," *Transactions A.F.S.*, **63**, 271, 1955.

5. Adams, C. M., Jr., and Taylor, H. F., "Fundamentals of Riser Behavior," *A.F.S. Transactions,* **61,** 686, 1953.

6. Adams, C. M., Jr., and Taylor, H. F., "A Simplified Analysis of Riser Treatments," *Transactions A.F.S.,* **60,** 617, 1952.

7. Pellini, W. S., *Solidification of Various Metals in Ingot and Sand Molds,* Electric Furnace Steel Conference, A.I.M.M.E., Chicago, 1956.

8. Bishop, H. F., and Ackerlind, C. G., "Dimensioning of Risers for Nodular Iron Castings," *Foundry,* **84,** 115, December 1956.

9. Caine, J. B., "Risering Castings," *Transactions A.F.S.,* **57,** 66, 1949.

10. Pellini, W. S., "Factors Which Determine Riser Adequacy and Feeding Range," *Transactions A.F.S.,* **61,** 61, 1953.

11. Bishop, H. F., and Pellini, W. S., "Solidification of Metals," *Foundry,* **80,** 87, February 1952.

12. Schmidt, W. A., and Taylor, H. F., "Risering of Gray Iron Castings, Progress Report No. 4," *Transactions A.F.S.,* **61,** 131, 1953.

13. Adams, C. M., Jr., Flemings, M. C., and Taylor, H. F., "Solidification and Risering of Gray Iron Castings," *Transactions A.F.S.,* **66,** 369, 1958.

14. Ruddle, R. W., *The Solidification of Castings,* Institute of Metals, London, 1957.

15. Dwyer, P., *Gates and Risers for Castings,* Cleveland, Penton Publishing Co., 1949.

16. Taylor, H. F., and Rominski, E. A., "Atmospheric Pressure and the Steel Casting—a New Technique in Gating and Risering," *Transactions A.F.S.,* **50,** 215–259, 1942.

CHAPTER 7

Gating

PART I. GATES AND GATING

7.1 The gating system

Gates are channels through which molten metal flows to fill a mold cavity. Their type and nomenclature depend upon the function they are designed to perform. The three main types are downgates, crossgates, and ingates, designed to lead the molten metal down through the mold, across the mold, and into the main mold cavity. In modern terminology, downgates, crossgates, and ingates are called *sprues, runners,* and *gates,* respectively. The complete assembly constitutes the *gating system.*

Figure 7.1 illustrates a simple type of gating system with modern nomenclature. Figures 7.2 through 7.6 show various modifications of the sprue-runner-gate arrangement. Gating systems such as these can be formed from prefabricated pattern pieces which are fastened to or separate from the patterns, or they can be cut directly in the mold with hand tools. Prefabricated systems are carefully engineered and have been used extensively on large castings for many years; they are being used to an increasing extent today for smaller castings.

The functions of an ideal gating system are to (1) fill the mold cavity, (2) introduce molten metal into the mold with as little turbulence as possible to prevent mold erosion and gas pickup, (3) establish the best possible temperature gradients in the casting, (4) introduce proper skimming action on the metal as it flows through the sprue system, and (5) regulate rate of entry of metal into the mold cavity.

In order for gates to function properly, one must control (1) rate of pouring, (2) size, number, and location of gates leading

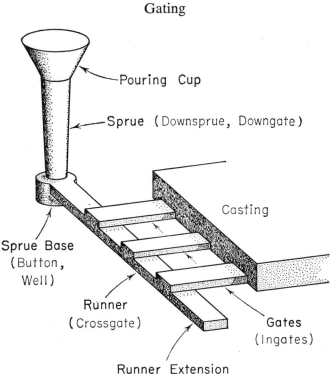

Pouring Cup

Sprue (Downsprue, Downgate)

Casting

Sprue Base
(Button,
Well)

Runner
(Crossgate)

Gates
(Ingates)

Runner Extension

FIG. 7.1.   Gating terminology.

to the casting, (3) size and type of sprue and runner, (4) type of pouring equipment, such as ladles, runner cups, or pouring basins, (5) position of the mold during casting and freezing, and (6) temperature (fluidity) of the metal.

7.2   Top gates

Top gates are usually limited to relatively small molds of simple design or to larger castings made in molds of erosion-resistant material.   The turbulence of the metal as it enters the cavity (Fig. 7.2) tends to erode portions of the mold, as well as to cause entrapment of air and metal oxides in the casting itself.   In the pouring of steel and cast iron, mold erosion is usually the most severe problem encountered from the turbulence of top gating; metal ingot molds and molds made from chamotte may be used to withstand this erosive action and permit pouring from con-

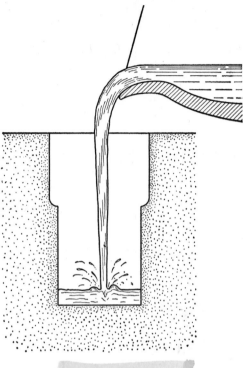

FIG. 7.2. Top gating.

siderable heights. In the lighter, more oxidizable metals such as aluminum and magnesium, drossing and air entrapment are the more severe problems resulting from turbulent pouring. Top gating of these metals is not recommended.

Pencil gates are often used for top pouring (Fig. 7.3). The gates control the rate of flow, and slag or dross inclusions tend to be cleaned from the metal in the pouring cup over the gates. Even with top pencil gates, however, the metal tends to enter the mold in severely turbulent fashion.

7.3 Parting line gates

Next to top gating, it is always easiest to place gates at a natural parting line of the mold (Fig. 7.4). Unless the parting line is at the bottom of the casting some turbulence occurs as

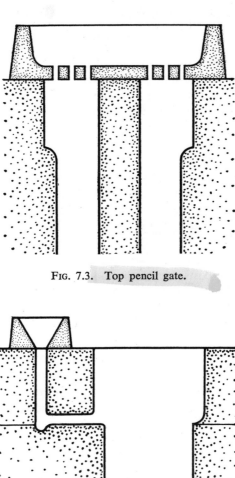

FIG. 7.3. Top pencil gate.

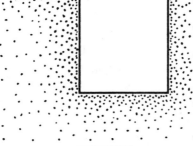

FIG. 7.4. Parting line gate.

the metal cascades into the mold cavity. From a fluid-flow standpoint, parting line gates are a compromise between top and bottom gates; they are often chosen more as a molding expedient than for their intrinsic value.

The turbulence resulting from metal cascading in the mold cavity may be minimized by designing the gating system so that the metal enters the cavity at a relatively low velocity and thereby tends to "dribble" down the mold walls rather than spurt into the cavity and impinge on the molds or core walls. A low entrance velocity is obtained by designing the gating system so that the gate cross section is considerably larger than the cross section of the sprue. Such a system is termed "sprue choked," or "unpressurized"; it is discussed in detail later in this chapter.

Parting line gates have the advantage that it is often easy to gate directly into risers. By so doing, the last (and hottest) metal poured enters the riser, promoting directional solidification. Cleaning costs are minimized by gating into risers since the gate entrance is not a separate appendage on the casting. Figures 7.4 and 7.6 illustrate types of "riser gating."

7.4 Bottom gates

Bottom gating reduces turbulence and erosion in the mold cavity to a minimum, but bottom gates may cause unfavorable temperature gradients when used with top risers (Fig. 7.5). The lower region of a bottom-gated mold is heated by metal flowing over it, and freezing is delayed in these locations. Care must be taken to assure that the potential "hot spot" at the gate entrance is minimized, or shrinkage may result. Overheating of the sand at any one location may be reduced by using a number of separate ingates. Excessive padding of casting sections toward risers, and use of extra large risers are sometimes also required to overcome unfavorable temperature gradients produced by bottom-gated castings having top risers.

Bottom gating is completely desirable when side risers are used. Hot metal is fed directly to the riser without first passing through the casting cavity, and the gate "hot spot" occurs in the riser where it promotes directional solidification (Fig. 7.6). Sometimes it is possible to use short *blind* risers (Fig. 6.6) to feed

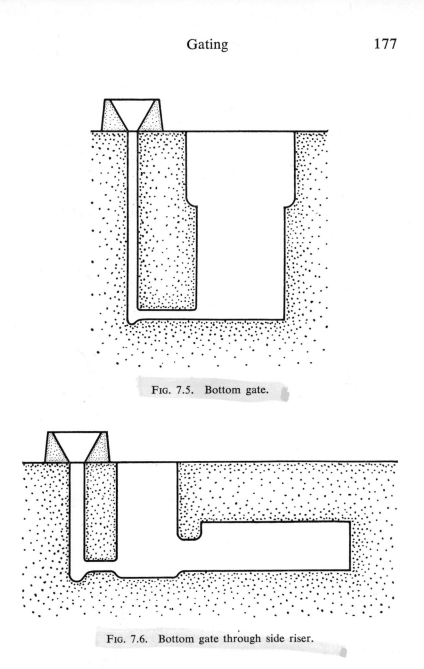

FIG. 7.5.   Bottom gate.

FIG. 7.6.   Bottom gate through side riser.

sections deep in the mold; gating through these risers is particularly desirable.

7.5   Mold manipulation

Sometimes bottom gating can be compatible with top risering by mold manipulation.   Figures 7.7 and 7.8 show how this is done.   Partial or total reversal of the mold after pouring places the hot metal above the cold and solidification progresses directionally toward the riser; if the system is designed and handled properly, hot molten metal is fed downward until all the casting is frozen.

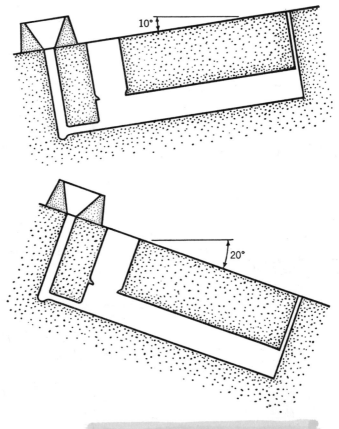

FIG. 7.7.   Mold manipulation; 30° partial reversal.

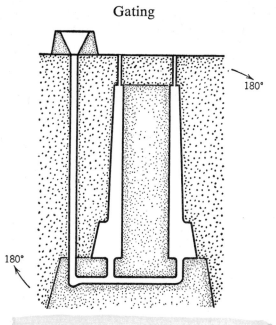

FIG. 7.8. Mold manipulation; total reversal.

7.6 Stepgates

Stepgates (Fig. 7.9) are designed to take advantage of the good features of bottom gating and correct the bad features. Metal flows through the bottom ingate until the mold is filled to the level of the next higher ingate; at this point metal is *expected* to flow through this ingate and through each successively higher one as the metal fills the mold. This *hypothetically* puts the hottest metal in the riser. Actually, stepgates do not function in this ideal manner; the inertia of the metal falling through the downgate carries it past the higher ingates and nearly all metal flows through the bottom one. By slanting the ingates upward at an angle to the casting, and designing the gates for relatively increasing resistance to flow at the lower levels, stepgates can be made to function properly. Unfortunately no formulas are available to aid in the design of stepgates; each must be selected by experience, common sense, and observations on the performance of various systems used for a particular metal from day to day. Size, type, and design of stepgates vary somewhat with casting size and metal temperature. One modification of the vertical gating system

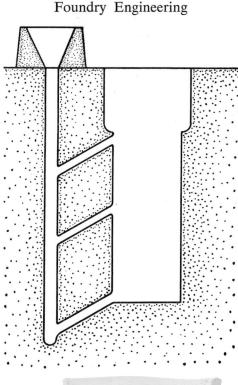

F<small>IG</small>. 7.9.   Stepgate.

which has been found to provide more uniform flow at successive levels is the "trombone" gate, illustrated in Fig. 3.3.

PART II.   PRACTICAL CONSIDERATIONS OF GATING DESIGN

7.7   Pouring cups and basins

Pouring cups and basins are used to (1) make it easier for the ladle operator to maintain the required flow rate, (2) minimize turbulence and vortexing at the sprue entrance, and (3) aid in separating dross and slag from the metal before it enters the runner system. Pouring cups (Fig. 7.10) serve only to make it easier for the ladle operator to direct the flow of metal, while pouring basins fulfill all the functions listed above.

Figure 7.11 illustrates some typical designs of pouring basins. These can be fabricated of core sand or of metal, or they can be cut or molded in the cope of the sand mold. The operator pours into the basin at the point farthest from the sprue hole; a dam is usually provided to enable the operator to reach optimum pouring speed before any metal enters the sprue. By not pouring directly into the sprue, danger of turbulence and vortexing with consequent air entrapment at the sprue entrance is minimized. Most pouring basins are large enough to permit some flotation of dross and slag so that only relatively "clean" metal enters the mold; the cleaning action is sometimes aided by placing a skimmer in the pouring basin (Fig. 7.11b).

7.8 Streamlined gating

Streamlining the gating system is always desirable from the standpoint of obtaining sound, clean castings. Often it is also advantageous from an economic standpoint; flow rates in gating systems may be so increased by streamlining that the various gating components may be decreased in size and a higher casting "yield" obtained.

Figures 7.12, 7.14, and 7.15 illustrate how improper gating design can result in metal turbulence and gas aspiration. These are the major effects which streamlining seeks to minimize since they may cause (1) entrapment of air or dross in the casting, (2) erosion of the mold walls of the gating system with consequent

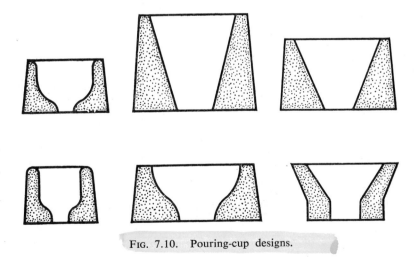

Fig. 7.10. Pouring-cup designs.

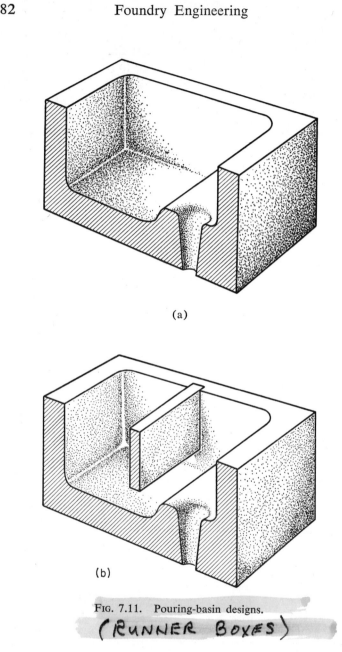

(a)

(b)

FIG. 7.11.   Pouring-basin designs.

(RUNNER BOXES)

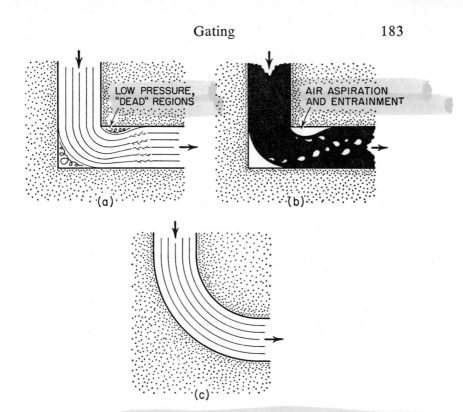

Fɪɢ. 7.12.   Fluid flow around bends in a gating system.   (a) Turbulence resulting from a sharp corner.   (b) Metal damage which occurs at the sharp corner.   (c) Streamlining the corner.

entrapment of sand in the casting, and (3) reduced metal flow rate.

Figure 7.12a shows the turbulence that occurs at a sharp corner in a gating system.   Low-pressure "dead" regions exist at the corner, and, if the metal velocity is sufficiently high, the pressure may actually drop below atmospheric in these regions.   When it does so, mold gases are drawn, or *aspirated,* from the permeable mold into the flowing stream (Fig. 7.12b).   Substitution of an adequate radius for the sharp corner minimizes the turbulence and eliminates aspiration.   The streamlining also results in a substantial increase in flow velocity (Fig. 7.13).

Effects similar to those at corners occur at sudden contractions

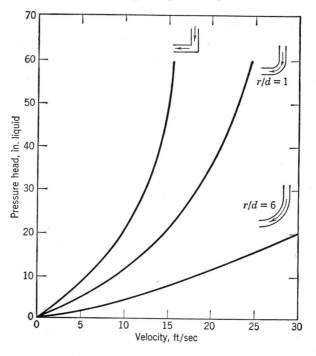

FIG. 7.13.   Effect of streamlining gating system on flow of metal.   (After Lips.)

or expansions of flow channels (Fig. 7.14).   Again, turbulence and possible air aspiration occur.

Figure 7.15 illustrates the necessity of thoroughly streamlining sprues to minimize turbulence.   Severe aspiration occurs in straight (uniform cross section) sprues; the aspiration results because the metal velocity increases as it descends the vertical sprue.   If the sprue is tapered by an amount sufficient to assure that the metal lies firmly against the mold walls, aspiration is reduced and turbulence minimized.   An additional precaution can be taken by radiusing the sprue entrance and exit to prevent deleterious "entrance effects" such as those shown in Fig. 7.14.

At the juncture between the sprue and runner, a gentle radius is difficult to mold in a vertical plane, and a "well" such as that of Fig. 7.15b is often used to absorb the kinetic energy and minimize aspiration.

7.9   The gating ratio

"Gating ratio" is the term used to describe the relative cross-sectional areas of the components of a gating system. It is usually defined as the cross-section ratio of sprue area: total runner area: total gate area. For example, a gating system which has a sprue of 1 sq in. cross section, a runner of 2 sq in., and two gates each of 1 sq in. cross section, has a gating ratio of 1:2:2.

Gating ratios recommended in the literature for foundry use vary over a wide range, but they may be grouped into two general classifications: *pressurized* and *unpressurized* systems. In the pressurized system a back pressure is maintained on the gating system by a fluid-flow restriction at the gates; this usually requires that the total gate area be not greater than the area of the sprue, as for example in systems with gating ratios of 1:0.75:0.5, or 1:2:1. In unpressurized systems, the primary restriction to fluid flow is at or very near the sprue; gating ratios such as 1:3:3 are used for this type system.

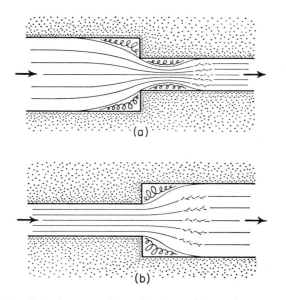

Fig. 7.14. Turbulence resulting from a sudden enlargement or contraction in a gating system.

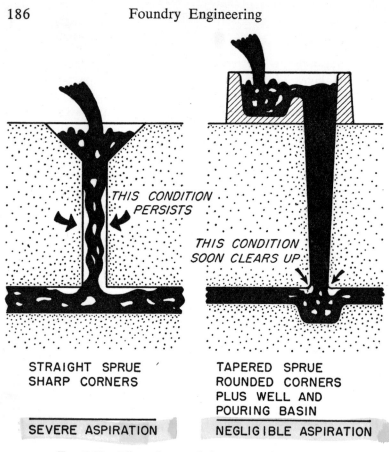

FIG. 7.15.   Effect of sprue design on metal turbulence

Figure 7.16 illustrates a typical pressurized gating system such as might be used for casting gray iron; the gating ratio is 1:0.75:0.50; no streamlining is shown.   Advantages of a properly designed system of this type are:

1. The gating system is kept full of metal.   The back pressure due to the restriction at the gates tends to minimize danger of the metal pulling away from the mold walls with consequent air aspiration; if the back pressure is sufficient, even untapered sprues may be kept full of metal (after the initial stages of pouring).

2. When multiple gates are used, flow from each of the gates (if the gates are of equal area) is about equal.   In unpressurized

systems, the kinetic energy of the metal stream tends to carry it down the length of the runner and preferentially out the gate farthest from the sprue. The restricted gate areas of the pressurized system tend to minimize these kinetic effects, and flow is more nearly equal from equal-size gates.

3. Pressurized systems are generally smaller in volume for a given metal flow rate than are unpressurized systems; thus less metal is left in the gating system and casting "yield" is higher.

Pressurized gating systems have certain disadvantages which result from the relatively high metal velocities in the flow channels. Severe turbulence may occur at junctions and corners unless careful streamlining is employed. Also the high velocities with which the metal enters the mold cavity may result in additional severe turbulence there, with consequent air entrapment, dross formation, and mold erosion. Present industrial practice favors the use of mildly pressurized systems for ferrous metals (especially for cast iron), where some metal turbulence in the gating system does not

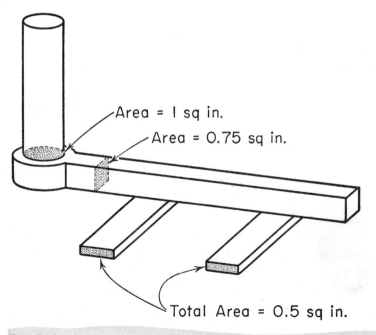

FIG. 7.16. A pressurized gating system (gating ratio—1:0.75:0.5).

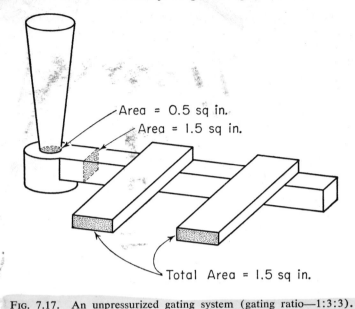

Area = 0.5 sq in.

Area = 1.5 sq in.

Total Area = 1.5 sq in.

Fig. 7.17.   An unpressurized gating system (gating ratio—1:3:3).

appear especially deleterious to metal quality; however, if metal velocity into the mold becomes too high, mold erosion can result. Brasses and bronzes are gated with either pressurized or unpressurized systems, depending on the individual shop and the alloy poured.   The light, oxidizable metals (aluminum and magnesium) are normally gated with unpressurized systems; in these alloys turbulence in the gating system must be kept to a minimum, and entrance velocities into the mold must be low to avoid air entrapment and dross formation.

Unpressurized systems have the advantages of lower metal velocity as compared with pressurized systems.   The large cross-sectional area of the runner and gates permits adequate flow rates at relatively low velocities.   Thus turbulence in the gating system and "spurting" of the metal into the mold cavity are reduced. Figure 7.17 illustrates a partially streamlined unpressurized gating system with a gating ratio of 1:3:3.

Disadvantages of unpressurized systems are:

1. Careful design is required to be sure that unpressurized systems are kept completely filled during pouring.   Since the gates

exert little or no back pressure, improperly designed sprues and runners may never fill completely or, if they do, separation effects such as those of Fig. 7.14 may easily occur. Drag runners and cope gates aid in maintaining a full runner, but careful streamlining is essential to eliminate the separation effects and consequent air aspiration. Tapered sprues with sprue wells are often employed. Streamlining of the junction between runner and gates is also recommended, with reduction in runner size at each gate as shown in Fig. 7.18.

2. It is difficult to obtain equal flow from multiple gates in unpressurized systems. As with stepgates, the kinetic energy of the flowing metal in unpressurized systems tends to carry it down the length of the runner and out the gates farthest from the sprue. Careful design, including reduction of runner size after each gate, can be used to obtain uniform flow through all gates but some trial and error is still usually necessary in arriving at proper gating ratios.

3. Unpressurized systems require large runners and gates. Because of the relatively large cross section of gates and runners, unpressurized systems generally are larger than pressurized systems of the same flow rate; thus casting yield is somewhat reduced.

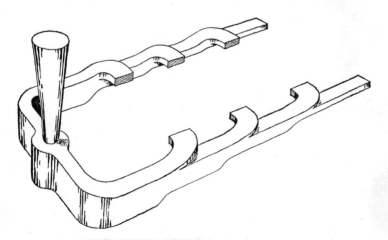

FIG. 7.18. A fully streamlined gating system. (After Grube and Eastwood.[2])

7.10   Eliminating slag and dross

For one or another reason even the most carefully designed gating systems may have slag, dross, or air entrapped in the metal flowing through it.   These foreign materials can originate in the melting furnace or the ladle and be carried into the gating system; they can arise from faulty pouring; or they can be produced in the gating system itself.   Especially in the early stages of pouring when the runner system is incompletely full, some air entrapment and dross formation is probably inevitable.

To prevent foreign materials in the flowing stream from reaching the casting, various steps are often taken to "clean" the metal as it passes through the runner system.   One of the most common of these is the runner extension (Fig. 7.1).   Since the first metal through a runner system is likely to be damaged by the initial turbulence, the purpose of the runner extension is to prevent this metal from reaching the mold cavity.   The first metal tends to be carried by its kinetic energy past the ingates and into the "dead end" of the runner; only the following, clean metal enters the casting.

Another technique of cleaning certain metals is to place the runner in the cope and gates in the drag, as sketched in Fig. 7.16. In this arrangement, foreign materials float to the surface of the flowing stream in the cope runner, while clean metal is tapped off the bottom of the runner to the casting.   Cope runners and drag gates are effective metal cleaners for (1) metals where the dross (or slag) floats easily to the surface and (2) where a pressurized gating system is used.   This type of system is, therefore, most widely used for ferrous metals and some copper-base alloys; it is not much used for light metals where dross does not readily float out and where pressurized systems are usually undesirable.

Metal cleaning by flotation is aided by using very large runners: i.e. as with a gating ratio of 1:4:1.   In this case, the reduced velocity in the runner permits more complete separation of the foreign particles in the runner.   Sometimes, also, serrated "teeth" are molded in the top of the runner to trap such particles.   The swirl gate (Fig. 7.19) employs centrifugal force to aid in the separation of slag.   In this gate, the lighter, foreign material is forced to the center of the swirl and up a riser or vent.

In light alloy casting, techniques such as those described above cannot be relied on to completely cleanse the flowing metal. When unpressurized, drag runner-cope gate systems are employed, the runners and gates are often made wide and shallow so that air and dross may float and adhere to the cope surface; also, runner extensions are employed to trap the first damaged metal. For complete metal cleaning, however, it is necessary to add screens to the runner system; perforated tin-plated sheet is normally used for this purpose.

Screens may be inserted in gating systems at the sprue base, in

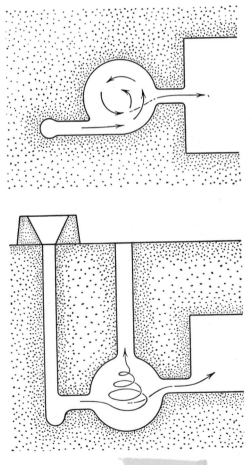

FIG. 7.19.   Swirl gate.

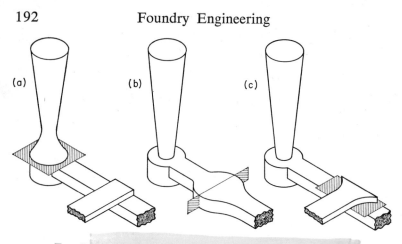

FIG. 7.20. Methods of placing screens in gating systems.

the runner or gates, or at the juncture between the runner and gates.   Figure 7.20 illustrates several possible arrangements. Wherever screens are located, it is usually desirable to enlarge the cross section of the flow channel at that point as otherwise the screen will act as a "choke."

In pouring relatively large magnesium castings, screens of the type shown in Fig. 7.20 tend to become plugged with dross.   To

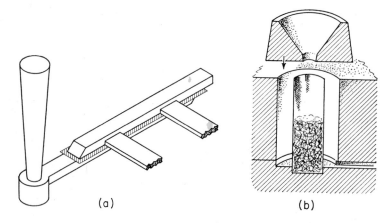

FIG. 7.21. Methods of placing screens in gating systems for severely drossing alloys. (a) Parting line screen. (b) Tubular screen in an over-sized sprue (screen is shown partially filled with steel wool).

avoid this, steel wool may be used as a filter in front of the screen, or a very large screen area may be achieved by use of techniques such as those illustrated in Fig. 7.21.

Strainer cores (disk-shaped sand and ceramic cores with small holes) are sometimes used in place of metal screens, especially for metals other than aluminum and magnesium. Because of the relatively large holes in these cores (compared with metal screens) it is doubtful whether they perform much actual filtering; the beneficial effects which are sometimes achieved by their use may be due to the fact the cores also act as a "choke." [3]

7.11 Gating for running and feeding

Heretofore, discussion has been limited to the design of gating systems that introduce clean metal into a mold cavity in a manner as free of turbulence as possible. To produce a sound casting, gates must also be designed to completely fill the mold cavity (prevent "misruns") and to promote feeding (establish proper temperature gradients).

Preventing casting "misruns" without use of excessively high pouring temperatures is still largely a matter of experience. To fill thin sections of castings completely, flow rates must be high, but not so high as to cause damaging turbulence. Multiple gates are frequently used, with individual gates extending to areas where a misrun is likely to occur. Adequate vents should be incorporated where a back pressure due to mold gases may otherwise occur and hinder metal flow. It should also be noted that many factors other than the gating arrangement and metal temperature may affect the ability of a molten alloy to fill a mold. These are the factors that affect metal *fluidity;* they include alloy analysis and gas content, and the heat-extractive power of the molding material. Often it is desirable to check metal fluidity before pouring with one of several *fluidity tests.*[4, 5] Figure 7.22 illustrates a standard fluidity spiral test widely used for cast steel. The "fluidity" of the alloy is rated as the distance, in inches, that the metal runs in the spiral channel.

The gating system must be designed to promote the best possible temperature gradients. Gating and risering are closely interdependent processes; for thin-section castings, the gate itself may even be designed to act as a riser. In larger castings, *riser gating*

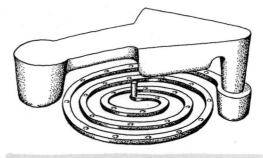

Fig. 7.22.   Fluidity spiral.   (After Taylor, Rominski, and Briggs.[4])

(Figs. 7.4, 7.6) is desirable to promote directional solidification toward the riser.   When it is necessary to gate directly into the casting itself, gates should be of relatively thin cross section so that they freeze off quickly and do not form an unfed "hot spot." Usually it pays to use *multiple gates* whenever practical.   They distribute and reduce hot spots, and they minimize penetration of metal into the sand near gates.   This is, of course, only a general rule and depends somewhat on the size and configuration of the casting.   If multiple side risers are used, it is best to gate through as many of them as possible in order to flow "hot metal" to each riser cavity.   Gating design should always be formulated to promote, insofar as possible, proper feeding of the alloy poured.

PART III.   THEORETICAL CONSIDERATIONS OF GATING

7.12   Introduction

This section will deal with the quantitative application of fluid-flow theory to gating practice.   To date, such applications in actual production of castings have been limited, partly because experience is still a convenient guide, and partly because gating systems used in practice are quite complex (from a fluid-flow standpoint). There is, however, no fundamental reason why the application of fluid-flow principles cannot be made more quantitative; i.e., why gating systems cannot be fully "engineered" from fundamental principles, and some excellent research has been carried out in recent years with this objective in mind.   As research data on metal flow in sand molds is assembled, hydrodynamic theory and gating practice will certainly find a larger common ground and, hopefully, the "cut and try" technique of gating design will be less necessary.

7.13   Turbulence in the gating system

In the classical hydrodynamic sense, the flow of liquid metals in gating systems is nearly always "turbulent." By this is meant that the individual metal atoms do not flow in straight "streamline" paths down the gating system, but travel from side to side as well as forward. At sufficiently low metal velocities, true "streamline" flow can be achieved, but these velocities are so small that it is nearly always impractical to design a gating system to obtain them.

It has been established that the flow of all fluids in ducts can be related by their Reynolds number, N_R:

$$N_R = \frac{\rho V d}{\mu} \qquad (1)$$

where N_R = Reynolds number, dimensionless
ρ = fluid density in pounds per cubic foot
V = flow velocity in feet per second
d = diameter of the duct in feet
μ = the viscosity of the fluid in foot-seconds per pound

When a liquid is flowing at such a velocity that the Reynolds number is less than about 2000, true streamline flow results. When the Reynolds number is over 2000, flow is usually turbulent although special precautions may permit streamline flow to occur at Reynolds numbers as high as about 4000.

Reynolds numbers of about 2000 to 20,000 are obtained in ordinary gating systems; flow in these systems is nearly always turbulent (in the classical sense). The degree of turbulence encountered in well-designed gating systems does not appear to be harmful to metal quality, although when it becomes excessive damage results from (1) rupture of the liquid metal skin with consequent gas entrapment and (2) mold erosion with consequent sand or dirt entrapment. In practice, the design of gating systems does not involve the elimination of metal turbulence, but rather its reduction to a point where it is not harmful.

7.14   Velocity calculations

Two basic fluid-flow equations are of interest in design of gating systems, to (1) calculate metal velocity and flow rates, and (2) to

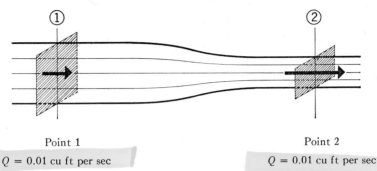

Point 1

$Q = 0.01$ cu ft per sec

$A_1 = 0.01$ sq ft

$V_1 = \dfrac{Q}{A_1} = 1$ ft per sec

Point 2

$Q = 0.01$ cu ft per sec

$A_2 = 0.005$ sq ft

$V_2 = \dfrac{Q}{A_2} = 2$ ft per sec

FIG. 7.23.   Effect of channel size on flow velocity. Q, A, and V are flow rate, cross-sectional area, and velocity, respectively.

obtain an understanding of the fundamentals of metal flow in gating systems. The first of these is the *law of continuity* which may be written:

$$Q = A_1 V_1 = A_2 V_2 \qquad (2)$$

where Q = metal flow rate in cubic feet per second
$\quad A_1$ = cross-sectional area of flow channel at point 1 in square feet
$\quad V_1$ = metal velocity at point 1 in feet per second

The law of continuity applies only to channels that are completely full; it states that, since liquids are incompressible, flow rate Q must be the same at a given time in all portions of a fluid system. Figure 7.23 illustrates use of the equation to predict flow velocities in a simple flow channel. If a metal is flowing at a rate, Q, of 0.01 cu ft per second in a flow channel of area, A_1, 0.01 sq ft, its velocity, V_1, is equal to Q/A_1, or 1 ft per second. If the flow channel then narrows to half its original cross section, the metal velocity must double.

The second equation of basic importance in flow calculations is *Bernoulli's theorem:*

$$\frac{V_1{}^2}{2g} + \frac{P_1}{\rho} + h_1 = \frac{V_2{}^2}{2g} + \frac{P_2}{\rho} + h_2 \qquad (3)$$

where V_1 = metal velocity at point 1 in feet per second

g = acceleration due to gravity, 32.2 ft per second per second

P_1 = static pressure in the liquid at point 1 in pounds per square feet

ρ = density of the liquid in pounds per cubic feet

h_1 = height of liquid at point 1 in feet

Bernoulli's theorem states that the energy of a liquid at a given point can be separated into three parts, energy of *velocity* ($V^2/2g$), energy of *pressure* (P_1/ρ), and energy of position (h). In the ideal case (no friction or other energy losses) when the liquid moves from point 1 to point 2, it neither gains nor loses energy. Thus, setting the energies equal for the two positions yields the equation above.

Figure 7.24 illustrates how Bernoulli's equation can be used to predict flow velocity in an *idealized* gating system; flow velocity is to be calculated at point 2, the point where the runner (or gate) enters the mold cavity. The height of the sprue is 1 ft, and in the idealized case no energy losses occur in the flowing stream. At point 1, the pressure is 1 atmosphere (since the metal is here exposed to the atmosphere), the velocity is zero (for the case of a large pouring basin), and the height is 1 ft above the runner. At point 2, the pressure is also 1 atmosphere (since the metal is entering the mold) and the height is arbitrarily taken as zero. Applying these constants to equation 7.3, the pressure and initial velocity terms drop out, so that $V_2 = \sqrt{2gh}$ and the calculated velocity V_2 is 8 ft per second.

If the exit area of the flow channel is known, the *flow rate Q* may be readily calculated. For example, assume the exit area to be 0.01 sq ft; then

$$Q = A_2 V_2$$

$$= (0.01)(8)$$

and $\qquad\qquad Q = 0.08$ cu ft per sec

The exit velocity and metal flow rate obtained by the above calculations will be somewhat higher than those found in actual practice; the reason is that in all flow channels some energy is lost when liquid flow passes from one point to another. The energy is lost in various ways including (1) resistance of the mold

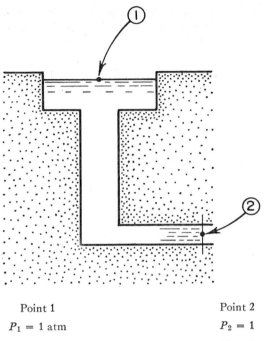

Point 1	Point 2
$P_1 = 1$ atm	$P_2 = 1$
$V_1 = 0$	$V_2 = ?$
$h_1 = 1$ ft	$h_2 = 0$

$$\frac{V_2^2}{2g} = h_1$$

$$V_2 = \sqrt{2gh}$$

$$V_2 = 8 \text{ ft per sec}$$

FIG. 7.24. Flow velocity calculation in an "ideal" sprue-runner system (no energy losses). P, V, and h are pressure, velocity, and metal head, respectively.

walls to the passage of metal (interface friction), and (2) internal friction (viscosity) of the metal; also playing an important part are the *entrance* effects and *bending* effects described earlier.

7.15   The tapered sprue

As metal descends a sprue, its velocity increases owing to the acceleration of gravity, and if the sprue is not tapered the metal pulls away from the mold walls with consequent turbulence and aspiration. If the walls are tapered sufficiently so the metal lies firmly against them during its fall, aspiration is eliminated. Equations 2 and 3 can be used to formulate an equation describing the taper necessary to prevent aspiration. The method for doing this may be found in the literature; [6] the result obtained is

$$\frac{A_1}{A_2} = \sqrt{\frac{Z_2}{Z_1}} \tag{4}$$

where A_1 = area of the sprue entrance

A_2 = area at any other location in the sprue

Z_2 = distance from top of the pouring basin to the location of A_2

Z_1 = level of the pouring basin above the sprue entrance

Equation 4 indicates that the ideal sprue should have a parabolic taper, but it has been suggested that in practice it is sufficient to calculate the inlet and outlet areas from the equation and draw a straight-sided taper between them.[6] Figure 7.25 illustrates these tapers. The solid line shows the parabolic taper and the dotted line shows the more practical straight taper. Equation 4 does not allow for entrance or exit effects; in practice the sprue entrance should be generously radiused. In other respects the taper

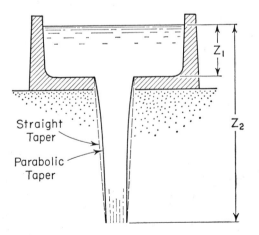

FIG. 7.25. Theoretical taper of sprues.

calculated from the equation is probably somewhat conservative, since energy losses in the sprue and beyond create a back pressure which reduces the amount of taper necessary.   Sometimes, in fact, the frictional energy loss in the sprue is intentionally increased by the use of rectangular cross-section or "slot"-shaped sprues.

7.16   Velocity calculations in real gating systems

It has been shown that Bernoulli's theorem can be employed to calculate velocities in ideal fluid systems: that is, in systems in which the fluid suffers no energy losses.   In real gating systems substantial energy losses occur at all channel entrances and exits, at bends, enlargements and contractions, and even in a smooth channel due to frictional effects.

Bernoulli's equation can be modified to account for these losses by the addition of appropriate terms to the equation.   In the case of friction in a smooth tube, the energy loss (per unit weight of metal) can be written

$$h_f = \frac{f \cdot L}{d} \left(\frac{V^2}{2g} \right) \qquad (5)$$

where h_f = energy loss due to friction in pounds
$\quad f$ = frictional loss coefficient, dimensionless
$\quad L$ = pipe length in feet
$\quad d$ = pipe diameter in feet
$\quad V$ = metal velocity through the tube in feet per second
$\quad g$ = acceleration due to gravity in feet per second per second

The energy loss at a bend (or pipe exit, etc.) can be written

$$h_b = K \left(\frac{V^2}{2g} \right) \qquad (6)$$

where h_b = energy loss at a bend or exit in pounds
$\quad K$ = loss coefficient at the bend or exit, dimensionless
$\quad V$ = velocity at the bend in feet per second
$\quad g$ = acceleration due to gravity in feet per second per second

These energy terms can be added to Bernoulli's equation in such a way as to describe the flow characteristics in any real system.

For example, to describe flow in a simple, uniform cross-section pipe with a single bend, Bernoulli's equation would be rewritten

$$\frac{P_1}{\rho} + h_1 = \frac{V^2}{2g}\left(K + \frac{f \cdot L}{d}\right) + \frac{P_2}{\rho} + h_2 \qquad (7)$$

Extensions of equation 5 to describe flow in more complex gating systems are readily accomplished. Such calculations are beyond the intent of this text, but they can be found in foundry literature,[3, 6] or developed from fluid-flow principles.

The use of equations such as these to calculate real flow velocities in actual gating systems is limited because of a lack of reliable data on the necessary loss coefficients (such as K and f in equation 7). Although some of these coefficients have been calculated, sufficient data have not been obtained for wide quantitative application of the principles. At the present time, it appears that these equations can best be used in practice to (1) determine the fluid-flow efficiency of existing gating designs, and (2) estimate the effect of possible changes in these gating designs on flow rate and on turbulence within the gating system.

7.17 Summary

Gating systems are composed of sprues, runners, and gates, to lead the metal down through the mold, across the mold, and into the mold cavity. In addition, pouring cups and basins are employed at the entrance to the gating system. Each of these components must be carefully engineered to assure that the gating system will (1) fill the mold cavity, (2) introduce molten metal into the mold with as little turbulence as possible to prevent mold erosion and gas pickup, (3) establish the best possible temperature gradients in the casting, (4) introduce proper skimming action on the metal as it flows through the sprue system, and (5) regulate rate of entry of metal into the mold cavity.

Castings may be *top-gated, parting-line-gated, bottom-gated,* or *step-gated.* Each of these gating systems has particular advantages and disadvantages with regard to fluid-flow characteristics and to obtaining optimum thermal gradients in the casting. Choice of one or another of the gating systems for an individual casting depends on (1) casting configuration, (2) metal poured, and (3) economic considerations.

Streamlining of gating systems is always desirable from the

standpoint of obtaining sound, clean castings. Often, flow rates in a gating system can be so increased that streamlining can be justified from an economic standpoint (the size of gates and other components can be reduced). Sections of the gating system most prone to turbulence and air aspiration are (1) sprue, (2) junction between the sprue and runner, (3) sharp comers in the runner or gates, and (4) sharp changes in section of the runner or gates. Proper application of fluid-flow principles can reduce turbulence in these areas to such a low value that it is not harmful to metal quality.

Gating systems may be *pressurized* (back pressure maintained on the system by fluid-flow restriction at the gates), or they may be *unpressurized* (primary restriction to fluid flow at or near the gates). Section 7.9 outlines the advantages and disadvantages of each type of system. Present industrial practice favors the use of mildly pressurized systems for ferrous metals (especially for cast iron), and unpressurized systems for aluminum and magnesium alloys. Brasses and bronzes are gated with either type, depending in part on the alloy poured.

Many techniques are available to clean slag or dross from metal as it flows through a gating system. These include use of screens or skimmers in the system, and careful control of size and location of runner and gates. Use of properly designed pouring basins can also aid in cleaning the metal.

In Part III of this chapter, fluid-flow theory has been discussed as it applies to the gating of metals. Most actual gating systems for castings are still designed empirically, but basic fluid-flow theory is finding greater application in the engineering of gating configurations. For example, the taper required to eliminate aspiration and turbulence in a sprue design can be readily estimated from basic constants. Also, metal flow rates through different gating designs can be calculated for cases where the necessary loss coefficients have been determined.

In the preceding two chapters, gating and risering have been considered more or less separately. However, it is well to reflect that gating and risering are (1) mutually dependent processes and (2) governed by rules which should never be compromised in the interest of expedience. Gates and risers for all castings should be determined by engineers specially trained for the job, and their selection should never be left to chance or the pet whims of anyone. An ounce of planning is worth a pound of guessing. Careful records should be kept of all changes in gating and risering so that progress will be recorded. As far as possible, gates and risers should be integral parts of each pattern.

7.18   Problems

1. Discuss briefly the functions of a sprue system. Make several typical and special sprue systems of simple design from plastic shapes, and pour water through them to observe differences in flow characteristics.

2. What factors must be considered in designing effective gating systems?

3. When are top gates used? Why are top gates often made as pencil gates? What factors limit the use of top gates? What is the purpose of strainer cores?

4. What makes parting line gates so desirable? Mention two possible disadvantages of such gates. What effects would these disadvantages have on the resulting casting?

5. Discuss the temperature gradients established in bottom-gated molds. Give three ways in which these gradients can be improved with respect to final casting soundness. List all reasons for using bottom gating and the good and bad effects of this method.

6. Why are tangential gates used? Would you recommend this type of gate for use with blind risers? Why? Use a sketch to illustrate your answer.

7. Why are stepgates used in the foundry? Are they usually successful in achieving their objective? Explain.

8. What is "gating ratio"? Why is it important?

9. Why are unpressurized gating systems usually used for casting aluminum alloys? What difficulties are encountered in designing unpressurized gating systems?

10. Why are mildly pressurized gating systems sometimes used for casting steel? What difficulties are encountered in their design?

11. When would you design your runner system with the runner in the cope and gates in the drag? When would you do the reverse?

12. What techniques would you use to clean slag from steel in pouring and gating? To clean dross from magnesium?

13. What factors affect metal "fluidity"? Why?

14. You are to cast a cylinder 4 in. in diameter by 8 in. high. The cylinder is molded entirely in the drag of a green sand flask; it is not risered and is top-gated. The cope of the flask is 10 in. high, and the

height of the metal during pouring (in the pouring basin) is 2 in. above the cope. A tapered sprue with a 1 sq in. exit area is employed; gating ratio is 1:3:3. Assuming no energy losses in the system, how long will it take to just fill the casting cavity?

15. Solve Problem 14, for the cylinder molded entirely in the cope, and bottom-gated, other conditions are the same.

16. What should be the entrance area to the sprue of Problem 14, to assure that aspiration does not occur?

17. Write to the U. S. Naval Research Laboratory, Anacostia, D. C., c/o Librarian, and request the loan of their films on gating practices; do the same to American Foundrymen's Society, Golf and Wolf Roads, Des Plaines, Ill., c/o Technical Director, and then arrange a showing to your classmates. These films will teach some things you will not appreciate otherwise and will give you worth-while ideas for theses and research programs.

18. When ingot molds are filled with molten steel this is done by pouring the metal directly down the open top, thus violating every rule or principle of good gating practice; these ingots often weigh 20 or 30 tons. List several ways you think might be used to improve this practice; then discuss each in detail with your instructor who will probably tell you that none of them can be done. Don't be discouraged—there must be a better way that is fully practical!

19. Large rolling-mill rolls (20 tons or more in weight) are made of cast steel, using sand molds, and a bottom-gated, tangential pouring system. All the while these molds are being filled, particles of sand are constantly "popping" off the sidewalls and raining down on the rising metal. Why? One speck of such sand trapped in the solid metal, if uncovered during machining and polishing of the roll, would cause rejection. Why doesn't the sand collect in the outside portions of the casting?

20. In the so-called German process of casting gun tubes centrifugally, a thin layer of unbonded sand is spun against the metal mold to a uniform depth. Why would this layer of loose, unbonded sand not erode as molten steel was poured over it to fill the mold? Also, since the density of steel is about three times that of sand, why would not the molten metal displace the sand grains and a bad casting result?

7.19 Reference reading

1. Lips, E. M., "Gating with Special Reference to the Optimum Flow Conditions in the Molten Metal," *Foundry Trade Journal,* **60**, 519–521, June 15, 1939.

2. Grube, K., and Eastwood, L. W., "A Study of the Principles of Gating," *Transactions A.F.S.,* **58**, p. 76, 1950.

3. Ruddle, R. W., *The Running and Gating of Sand Castings,* Institute of Metals, London, 1956.

4. Taylor, H., Rominski, E., and Briggs, C. W., "The Fluidity of Ingot Iron, and Carbon and Alloy Steels," *Transactions A.F.S.,* **49,** 9–93, 1941.

5. Krynitsky, A. I., "Fluidity Testing of Molten Metals," *Transactions A.F.S.,* **61,** 399, 1953.

6. "Symposium on Principles of Gating," *Transactions A.F.S.,* **59,** 1951.

CHAPTER 8

The Ingot as a Casting

In the preceding chapters, the casting process has been considered as it applies to objects made as nearly to the shape in which they are to be used as possible. *Ingots* are castings made in relatively simple shapes, and designed for subsequent fabrication by working (rolling, forging, etc.). Ingots are, nonetheless, castings in every sense of the word; most of the problems associated with their production are typical foundry problems. Also, certain defects in wrought materials (such as laps and seams), and disadvantages of some wrought alloys (such as poor properties in the short transverse direction) can often be traced to inadequate foundry controls in the casting of the ingot.

8.1 Introduction

Several different basic ingot types are shown in Fig. 8.1. Figure 8.2 shows the cross section of a typical ingot, complete with mold, stool, pouring plug, and hot top. Molds into which these ingots are cast are usually made of metal (cast iron, steel, or copper); seldom is an ingot of this type made in a sand mold. So-called *pigs* (ingots of pig iron and non-ferrous metals) designed for remelting purposes are made in large quantity in both sand and metal molds; their shape and size vary widely and are not particularly important considerations since the pigs only need be in convenient form for handling as melting stock. These are not included in the subsequent discussion; only ingots designed for fabrication by working will be considered at present.

Ingot *molds* may be made with thin water-cooled walls, but nearly all are of massive construction and depend upon the mass, thermal conductivity, and specific heat of the mold to chill the cast ingot at the proper rate. Block graphite has proved to be an effective mold material and may be used more extensively in the future. Thermal conductivity of graphite is roughly three times that of steel, and its density is about one-third; molds must be large in order to absorb an adequate amount of heat. Graphite is rela-

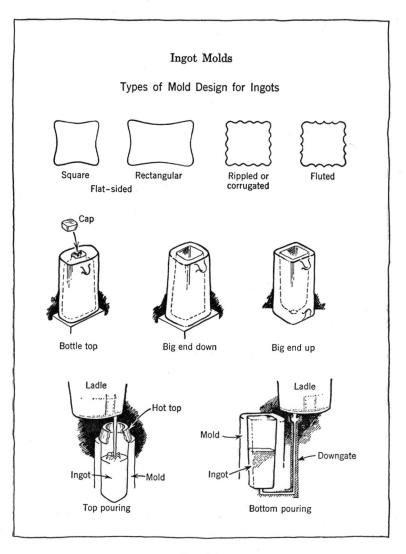

FIG. 8.1.

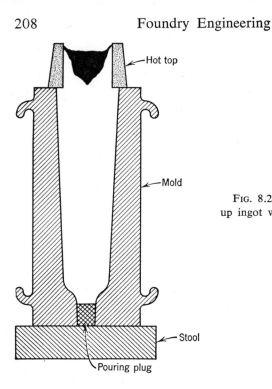

Fig. 8.2. Shrinkage in big-end-up ingot with hot top.

tively fragile and most suitable for small ingot molds; it has the advantage of not heat checking at the mold-ingot interface as readily as metal.

Current ingot casting practice has been developed empirically over many years. Plant equipment, often designed according to custom, expediency, or empirical observation, represents a great investment, and every technological improvement must be weighed carefully against economy of operation, cost of new facilities, and scrapping of old equipment; accordingly, practice is seldom altered except where ingot size and total plant change are relatively small.

An ingot represents outwardly the simplest casting possible; it has uniform, regular configuration without undercuts or protuberances; it requires no complicated cores; the mold is used repeatedly, and in general the metal is poured directly into the mold without special gates or runners. In spite of *apparent* simplicity, ingot casting is difficult; many imperfections of ingots are due directly to the simplicity of the casting process. Because of the engineering importance and relatively great accumulation of published knowledge, steel ingots will be discussed in detail in preference to ingots of other metals.

Steel ingots are divided according to internal structure into *rimming, semi-killed,* and *killed* types; externally they must be

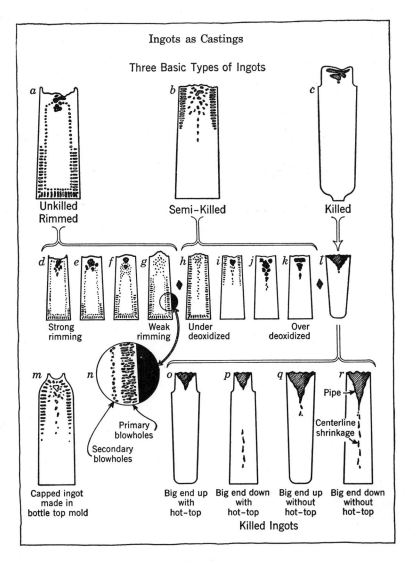

FIG. 8.3.

identified as *flat-sided square or rectangular, fluted, rippled or corrugated,* and *bottle-neck;* according to their method of casting, they are either *big-end-up* or *big-end-down,* and are either *top* or *bottom* poured. These characteristic types are shown schematically in Fig. 8.3.

8.2 Molds

Ingot *molds* are usually made of cupola or blast-furnace iron; the approximate composition is 3.5 per cent carbon, 1 per cent silicon, 0.9 per cent manganese, 0.2 per cent phosphorus, and 0.07 per cent sulfur. Molds must be heavy to extract enough heat from large ingots to prevent welding at the interface and to freeze the ingot at the proper rate so that an outside skin quickly forms and pulls away from the mold wall. Mold design is very important; flat-sided molds are ideal for rolling, but rippled or corrugated molds are sometimes needed to increase freezing rate through their relatively greater surface area. Often ingots crack and tear unless cooled rapidly; the fluted mold also provides a crystal structure less prone to cracking during initial forging operations. Large ingots must be tapered liberally and mold surfaces kept clean to prevent transverse tearing of the ingot as it settles in the mold. Tearing of ingots must be prevented at all cost.

8.3 Ingot structures

A *rimmed* ingot is one made from *underdeoxidized* steel (other shop terms with the same meaning are *effervescent, unkilled, rimming,* and *wild*), i.e., steel which continues to evolve carbon monoxide gas in the mold after pouring. The state of deoxidation of a steel is gaged respectively in the order of increasing completeness by the shop terms unkilled, semi-killed, and killed. The term rimmed is developed from the orderly disposition of *blowholes* along the outer rim of unkilled ingots (Fig. 8.3); ingot quality is affected by the size, amount, and distribution of these *primary* blowholes, and particularly by their depth beneath the thin, dense, outer skin of the ingot. The holes must *never* break through the outer skin, either during freezing or during the later heating and working operations, as the surfaces of oxidized discontinuities of any kind cannot be welded by hot working; the only reason blowholes can be tolerated at all in ingots is that they can be closed and welded into a homogeneous mass by drastic mechanical working. Obviously, unkilled steel can *never* be tolerated in *castings,* where no mechanical working is done, nor in fact in some ingots where mechanical working is relatively slight.

Figures 8.3*a*, *b*, and *c* show typical internal structures desired in rimmed, semi-killed, and killed ingots. Usually all three types are top-poured if ingots are large; the sketch indicates that big-end-up molds are used predominantly for killed steels, and big-end-down molds for others. The physical metallurgy, thermodynamics, and chemistry involved in explaining the rimming action of steel and the exact mechanism of blowhole formation are treated voluminously in technical literature, but only brief treatment is required for a working appreciation of the general phenomena most important to metal processing. Ingot steels may vary widely in alloy content, but all fall into the above three general categories.

8.4 Steel melting (general)

Steel for ingots is nearly always made in *open-hearth furnaces;* electric-arc and induction furnaces are used for making killed steels of special composition, and Bessemer or Tropenas converters are employed for making a relatively small tonnage of unkilled steels. Furnaces may be classed as *acid* or *basic,* depending upon characteristics of refractories used. *Steelmaking* is the process of refining impure scrap and pig iron by adding elements or compounds to, or taking them from, a molten bath of metal. *Slags* are used to protect molten metal from *uncontrolled oxidation* and as a *medium* for *transfer of elements* to and from the bath; slags are essentially *low-melting lime-silica* combinations in which the proportions of lime and silica determine whether the process and the steel are acid or basic. The most efficient unit for refining or purifying steel commercially is the *basic open-hearth furnace;* acid furnaces and converters are essentially melting units. The practical importance of the basic furnace lies not only in excellence of steel produced, but also in the fact that relatively poor raw materials may be used as melting stock; in other furnaces high-grade scrap is needed to produce good steel. This was a strategic consideration in World War II, when availability of good melting scrap was at an all-time low. Steel melting is considered in detail in Chapter 10.

8.5   Explanation of ingot structures and internal characteristics

The elements responsible for degree of rimming, or conversely degree of deoxidation, are carbon and oxygen; they may or may not combine into carbon monoxide gas.   The amount of carbon present in the molten bath (as it is poured from the furnace into ladles for transfer to the molds) can be controlled in the furnace during the refining process; the oxygen content is proportional to the carbon level (Fig. 8.4).   For a given carbon content oxygen can be controlled by addition of *deoxidizers* which react with some or all of the oxygen to form insoluble oxides.   A typical reaction is $2Al + 3FeO = Al_2O_3 + 3Fe$.   Such deoxidizers, in order of increasing effectiveness, are manganese, silicon, and aluminum. By analyzing quickly for carbon content of the bath, an approximation can be made of oxygen present and *deoxidizers can be added in the furnace ladle or ingot to give rimmed, semi-killed, or killed ingots,* as desired.   Generally, the amounts of deoxidizers used are determined empirically from experience.

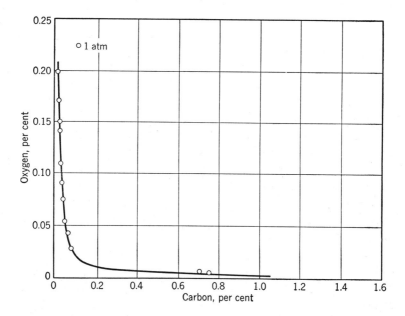

Fig. 8.4.   Effect of carbon content on solubility of oxygen in steel.

Blowholes develop near the ingot surface. A skin of relatively pure iron (low in carbon) solidifies quickly at the mold-metal interface and oxygen is rejected from solution; as oxygen concentrates ahead of the solid wall, the carbon-oxygen concentration product increases until gas bubbles form. Many bubbles rise from the bath, but some will be trapped by the advancing wall of solidifying metal. At a certain critical value of freezing rate, ferrostatic head, and carbon-oxygen content, the primary blowholes form; the size, shape, distribution, and depth of primary blowholes beneath the ingot skin determine ingot quality, and much work has been done to develop a scientific explanation of the process to make possible closer control of ingot structure. The region of secondary blowholes shown in Fig. 8.3*n* does not affect ingot quality as much as the primary zone.

Besides using aluminum for adjusting rimming action, ingots are sometimes *capped* at some definite time after pouring; this increases pressure on the metal as gas bubbles continue to form and ultimately inhibits their formation. Capping an ingot consists only in sealing the entire top surface of the metal in the mold with a concave metal cover; special *bottle-top* ingot molds are used to facilitate capping (Fig. 8.3*m*).

Bubble formation is a vital consideration in rimmed steels, of somewhat less critical importance in semi-killed steels, and of no significance in killed steels, where bubbles are never allowed to form. Other considerations of importance are best illustrated by studying the solidification of *killed*-steel ingots.

In steelmaking, close attention is given to controlling *carbon, manganese, silicon,* and *phosphorus*—basic elements common to all steels. The behavior of these elements, particularly their segregation during solidification, and the mechanism of solidification of metals in general, must be well understood to appreciate intrinsic properties of metal structures, whether the metal structures have been worked to shape, welded, or used as-cast. A repetition of general principles is worth while.

Macroscale features of a *killed*-steel ingot are shown schematically in Fig. 8.3*c*; no blowholes are present, but a large shrinkage cavity, or *pipe,* has formed at the top of the ingot. This so-called pipe is a troublesome manifestation of the decreased volume (or conversely the increased density) which accompanies the change

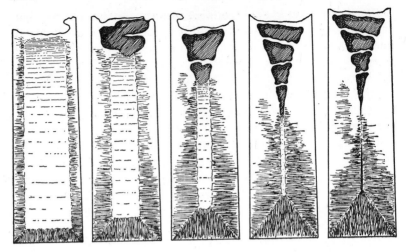

Fig. 8.5. Ingot sections of stearin during successive stages of solidification. (After Brearley.)

of most materials from liquid to solid state. The pipe develops as a result of *liquid* and *solidification* contraction.

Stearin was used by Brearley to simulate solidification and shrinkage behavior of gas-free metals. Molten stearin was poured into typical small, big-end-down ingots and allowed to solidify for different periods; the ingots were bled, cooled, and sectioned. Figure 8.5 shows sketches from Brearley's actual experiments. Solidification proceeds by the growth of crystals of stearin inwardly, roughly perpendicularly to the ingot wall, into the unsolidified stearin at the center. These columnar crystals meet in *pyramid-shaped* surfaces of demarcation where crystals meet as they grow simultaneously from the sidewalls and bottom of the mold.

These same phenomena occur in metal ingots to somewhat lesser degree; the planes where walls of advancing metal dendrites meet are definite planes of weakness in steel ingots, and special processing is often necessary to prevent cracking during initial *breakdown* of the ingot by working. The slope of the sidewalls of the cavity shows that solidification progresses steadily; thin layers of solid metal often *bridge* across shrinkage cavities, forming what is popularly called *secondary* piping.

Steel undergoes a volume contraction in cooling from the pouring temperature to the point of solidification of about 0.9 per cent for each 100°F the steel is above this temperature (degrees superheat). Also, as the metal changes from liquid to solid, its density increases approximately 3 per cent. If poured at 360°F (200°C) superheat, total contraction would be roughly 6.2 per cent of the original volume, and in a large ingot the pipe would clearly be of considerable size. If the walls of the ingot mold are parallel, an elongated pipe may extend down the center of the ingot nearly to the bottom; this low-density area is called *center-line* shrinkage. In order to minimize the seriousness of the shrinkage by localizing it at the top, molds with tapered walls of the big-end-up type are used in conjunction with a hot-top or insulated riser; this delays solidification of the metal at the top until after the ingot is completely frozen. Molten metal is thereby provided in the riser (hot-top) at all times during solidification of the ingot.

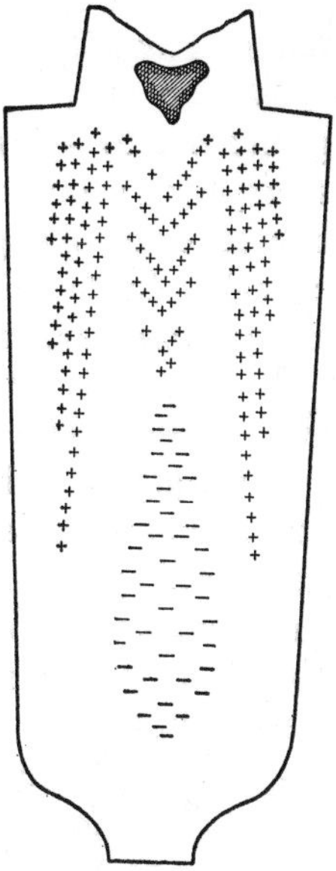

Fig. 8.6. Characteristic positive and negative segregation in a killed-steel ingot.

8.6   Segregation in steel ingots

Segregation of elements and impurities occurs in ingots as it does in sand castings; this is a condition in which areas of the ingot are either richer or poorer in certain constituents than the average composition. Figure 8.6 shows the general location and configuration of zones of *positive* and *negative* segregation in a killed-steel ingot. The zone of *negative* segregation is caused by settling of the relatively pure crystals as they nucleate freely in the un-altered melt during late stages of solidification. This is one form of so-called *macrosegregation.* *Positive* segregation, another form of *macrosegregation,* exists because the falling crystals displace liquid metal in a rising current; this process, with freezing con-tinuing steadily inward from the mold walls, produces the V and inverted V zones of positive segregation. Macrosegregation is greatest in rimming steel ingots owing to the effervescent rimming action; it is least in killed steel ingots.

Composition changes often occur over areas too small to be subjected to chemical analysis. The process of freezing is, of course, selective, and all stranger atoms tend to segregate. Some elements like carbon diffuse rapidly through the solid matrix, and the temporary segregation is readily wiped out; manganese and sulfur diffuse less readily, and localized *microsegregation* exists, particularly at boundaries of columnar grains. This segregate pre-vails even after metal has been heated and worked extensively, giving rise to a *banded* structure under the microscope. Micro-segregation is due to enrichment of so-called stranger atoms in the interdendritic fillings and their subsequent entrapment between crystals. In steel, the elements sulfur, oxygen, carbon, phosphorus, silicon, and manganese, in that order, tend to segregate least.

8.7   Semi-killed ingots

Rimming steel ingots are characterized by primary and sec-ondary blowholes and by complete absence of piping and center-line shrinkage; in killed steels, the exact opposite is true: piping and sometimes center-line shrinkage exist and there are no blow-holes. It has been learned that blowholes are not harmful if they do not reach the surface and hence are not oxidized, and if the ingot is extensively worked so that holes are flattened and welded. Shrinkage pipes in ingots represent waste to the mill operator, as metal in the vicinity of the cavity must be cut from the useful part of the ingot and discarded. An attempt to combine the useful features of killed and rimming ingots is represented by semi-killed steel ingots. In these, molten metal is partially deox-idized and the carbon-oxygen ratio is adjusted so that only mild gas evolution develops. The interplay of shrinkage, which reduces pressure and allows carbon monoxide to form more easily, and formation of gas bubbles which reduces shrinkage, result in ingots such as shown in *b, h, i, j,* and *k* of Fig. 8.3.

8.8 Non-metallic impurities

Cast steel always contains solid non-metallic impurities to a degree which depends upon the steelmaking and pouring practices, upon composition of the particular steel, and upon method of deoxidation; if the casting is subsequently worked, as in an ingot, these inclusions are pressed and drawn into stringers whose nature, amount, and particularly whose shape and distribution, determine the quality of sheet, plate, or other wrought shape. Not all such impurities are deleterious; sulfur, for example, is sometimes added to improve machinability. Some are purposely provided as the lesser of two evils. For example, manganese is particularly useful because of its great affinity for sulfur; in the absence of manganese, iron sulfide forms in a continuous film of eutectic at grain boundaries, melts at hot working temperatures, and causes *hot shortness.*

Non-metallic inclusions may be *indigenous* or *exogenous;* i.e., they may precipitate from reactions within the metal or they may be particles of slag and/or refractories washed in with the flowing metal. *Indigenous* impurities, or inclusions, may form as a result of oxidation, deoxidation, change in solubility or equilibrium constants with cooling, or enrichment of liquid by segregation. Inclusions may freeze before the metal, in which event they are globular in form, or as stringers of eutectics at crystal boundaries; pseudo forms combining both shapes are sometimes present also. Complex manganese silicates are examples of globular inclusions, while many sulfides form by precipitation from interdendritic mother liquor enriched in sulfur by segregation. Iron-manganese silicates, iron-silicon aluminates, oxides of iron, manganese, silicon, titanium, and aluminum, and sulfides of iron and manganese are a few examples of indigenous inclusions.

Exogenous impurities could be prevented in large part by closer attention to pouring (teeming) practices than is usual in ingot manufacture. Top-pouring of ingots relies entirely on such control as can be obtained with bottom-pouring ladles. Though most exogenous impurities rise to the top of the metal in the ladle by levitation, ample opportunity remains for erosion of the stopper rod and nozzle. Elaborate gating practices in making sand castings, described in a later chapter, are designed to provide additional control over this source of inclusion. Unfortunately, this luxury is too expensive for any but the highest-priced, killed ingots; tool steel and other special ingots are often bottom-poured with as many as six molds attached to a single downgate. In top-pouring, metal falls unprotected through air from considerable height. This encourages oxidation and causes splashing and turbulence. Many patents have been granted on objects and methods designed to reduce splashing at the bottom of ingots, but in general the only expedients used to obtain clean ingots are mold washes, initial *soft-pouring,* and rounded stools, which form a pool of steel quickly at the bottom of the ingot mold.

8.9   General considerations of steel ingots

Ingot type, for example rimmed, semi-killed, or killed, is dictated by specific considerations. (1) Rimmed steel is usually limited to carbon and manganese contents below 0.30 and 0.60 per cent respectively. (2) Economic considerations are principally furnace time, efficiency of alloy additions, and the yield of finished ingot. (3) Mill equipment determines whether ingots can be worked drastically enough to produce a dense product from unkilled steels. (4) Application, for example whether the steel is to be used as sheet, plate, strip, rod, wire, skelp, structural sections, rails, seamless tubing, or forgings, is also important. (5) Customer requirements for surface quality, mechanical properties, freedom from segregation, weldability, and final analysis must be considered.

The advantages and disadvantages of the three types of steel ingots are reasonably clearcut. Among the advantages of rimmed ingots are: (1) Surface finish is best because deoxidation products are a minimum, and the effervescing action discourages entrapment of oxide scum in the surface layers; killed steels are drastically deoxidized and any trapped impurity remains in place. (2) Proper rimming action provides a rim of clean, sound metal encasing the primary blowholes; this rim is very ductile and well adapted to deep drawing operations. (3) Piping is either eliminated, as in higher-carbon varieties of rimmed ingots, or relatively very limited as in the lower-carbon compositions.

The disadvantages of rimmed ingots include: (1) Maximum

segregation of metalloids from surface to center as consequence of their being continually swept from the liquid-solid interface by rimming action; the extent of this segregation depends upon how long effervescence is maintained. (2) Primary blowholes are not always controlled effectively and may break through the ingot skin and become oxidized. (3) Composition is limited by the carbon and manganese levels required to give adequate gas evolution, and to those elements which do not oxidize preferentially to iron: for example, nickel, copper, molybdenum, and phosphorus can be used, but chromium, etc., cannot.

Semi-killed steel ingots enjoy (1) higher yield than either rimmed or killed; (2) less segregation of metalloids than rimmed ingots; (3) lower residual deoxidizers than killed steels; and (4) a favorable economy, since furnace times are shorter than for killed steel and removal of pipe cavity is avoided. They suffer from (1) poor surface; (2) need for very close control of deoxidizing practice; (3) such blowholes as do exist must be welded and thus the degree of working is considerable; and (4) alloys are limited because of their effect on rimming behavior.

Ingots of killed steels (1) have an initially homogeneous structure which does not require drastic working; (2) can be made to closer chemical specifications of oxidizable and alloying elements; and (3) have minimum segregation—properties are uniform and high. The disadvantages of dead-killed ingots are that (1) production requires careful furnace practice; (2) shrinkage pipes are extensive and yields correspondingly low; (3) surface finish is relatively poor; (4) excessive deoxidizers are required to prevent rimming both during pouring and during solidification; and (5) hot-tops and special feeding must be provided.

Killed-steel ingots are used whenever structures made from them must be dense and homogeneous, when the amount of working is not drastic enough to completely eliminate the porosity of unkilled ingots, and also when chemical specifications for alloy and special steels do not lie within the allowable limits for rimmed or semi-killed steels. Forging and carburizing steels, stainless, special alloy, and tool steels are all dead-killed types.

Rimming steels are used whenever specifications allow up to 0.3 per cent carbon and 0.60 per cent manganese, whenever highly oxidizable elements are not required, and whenever subsequent working is drastic enough to homogenize porosity. Semi-killed steels are limited mainly to rolling into sheet, bar, skelp, plate, and structural sections.

Metalloids, metallic segregates, and any unclosed blowholes are elongated in the direction of rolling during the working of ingots. As a result of this (and also as a result of preferred grain orientation in the direction of rolling), mechanical properties in the transverse direction of an ingot are usually lower than in the direction of rolling. If, for example, ingot casting practice allows an excess of metalloids in the cast metal, transverse properties may be very low, and the metal from the ingot will be unsuitable for highly stressed applications. Fatigue life and impact strength are particularly sensitive to the effects of elongated metalloids and segregates. *Vacuum pouring* is a new technique whereby the metal for steel ingots is melted by normal practices and then poured *through* a vacuum into an ingot mold in a vacuum chamber. Degassing is accomplished during pouring and non-metallic impurities in ingots cast by this method are very low. As a result, fatigue life and other properties are improved and made more consistent. Ingots made by vacuum pouring are finding use for such applications as ball-bearing manufacture where fatigue life is of utmost importance. To date, ingots up to 100 tons in size have been cast in the United States by this method.

8.10   Non-ferrous ingots

Non-ferrous ingots are also of great industrial significance and fortunately are somewhat less complex in nature than steel ingots. Figure 8.7 shows a few typical ingot molds.

Ingots of lead, tin, zinc, aluminum, magnesium, copper, and pig iron destined for remelting purposes are cast into traylike, open-topped molds of copper, iron, or sand (Fig. 8.7a). Ingot molds for copper wire bar and large copper slabs are also of this type (Fig. 8.7b). The horizontal, open-top mold is generally not satisfactory for ingots to be worked into shapes, as the pipe, dross, and inclusions which mark the upper surface must be removed by machining; it is more satisfactory to cast in a mold with the long axis vertical so that the upper, exposed surface is minimized. It is sometimes necessary when casting oxidizable metals, or those that segregate badly (aluminum ingots, for example), to reduce turbulence and contact with air as much as possible and to pour the metal as slowly as possible so that freezing almost keeps pace with pouring. Figures 8.7c, d, and e are attempts to do this; the tilting mold reduces turbulence by permitting the ladle to empty metal close to and upon a slanting surface rather than dump the metal violently upon the bottom of

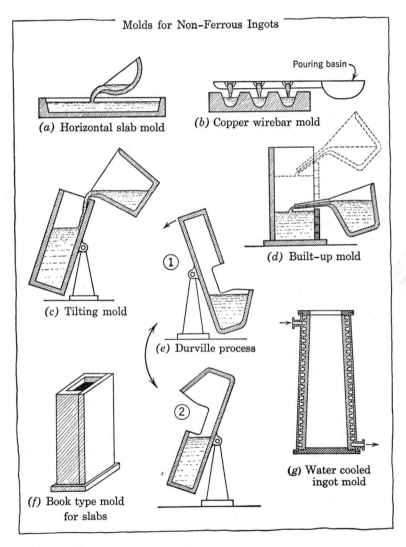

Molds for Non-Ferrous Ingots

(a) Horizontal slab mold

(b) Copper wirebar mold

Pouring basin

(c) Tilting mold

①

(e) Durville process

②

(d) Built–up mold

(f) Book type mold for slabs

(g) Water cooled ingot mold

Fig. 8.7.

a vertical mold. The built-up mold allows pouring close to the free upper surface of the metal by adding plates successively as the level increases. In the Durville method, metal is poured first into the cuplike container where dross and impurities remain when the mold is gently inverted. Split or book-type molds (Fig. 8.7*f*) are used extensively for casting flat billets of copper for drawing into shell cases and for similar operations. Water-cooled copper molds (Fig. 8.7*g*) are used for making brass strip ingots. Copper plates $\frac{1}{4}$ to $\frac{1}{2}$ in. thick with baffles are enclosed in a steel or cast-iron jacket, and water is circulated at a controlled rate. The mold is of two-piece construction, split vertically.

The only non-ferrous ingot purposely made to correspond to a rimming ingot of steel is *pitch copper* for working into wire bar and slabs; the copper is highly oxidized (usually to remove sulfur) and is then *poled* with a green wood stick to remove all but about 0.03 per cent oxygen; the hydrogen introduced by the green (undried) stick is objectionable only in special instances where working or performance requirements are drastic.

8.11   Special ingot-casting processes

Hollow ingots for tube drawing are sometimes required of metals that cannot be pierced easily; aluminum alloys are hard to pierce and stick to the mandrel, and leaded bronzes cannot stand heating for piercing; some steels are simply too hard or tough to be pierced economically.   Such ingots are made by providing ordinary vertical ingot molds with suitable cores or by centrifugal casting.

Many attempts have been made to make billets, strip, sheet, rod, and cylindrical ingots of aluminum, brass, and steel by a continuous process in which water-cooled, bottomless dies or rolls take the molten metal from ladle or furnace and convert it directly to semifinished form; i.e., the metal does not need to go through the ingot stage. Continuous casting of aluminum, brass, and steel rod is now a commercial process, and mills have been installed for horizontal, direct rolling of aluminum, brass, and steel strip.

8.12   Continuous casting

Casting round ingots, rods, slabs, square billets, and sheets by a continuous process directly from molten metal has been developed sporadically since the original patents of Laing, an American, in

1843, and Bessemer, an Englishman, in 1858. Such forms have been successfully cast from brass, copper, aluminum, and magnesium; continuous casting of plain high- and low-carbon steel, and of alloy and stainless steels, in billets 4 in. square, in slabs 4 by

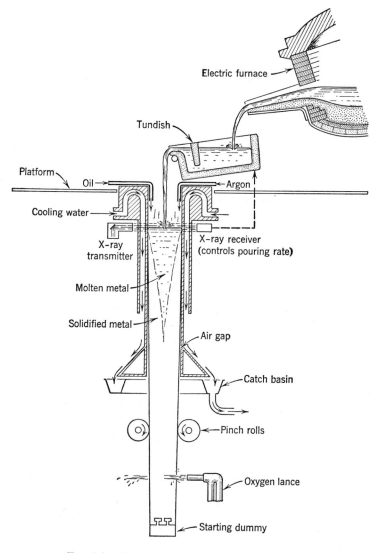

Fig. 8.8. Continuous method for casting steel.

$8\frac{1}{2}$ in., and in roughly 6-in. oval ingot, is a commercial reality. Aluminum ingots 8 by 8 and 18 by 20 in. are being made. Copper wire bar has been manufactured by this method commercially for several years.

In this process, molten metal is poured from a special ladle, called a *tundish,* into the top of a bottomless graphite or water-cooled metal mold of the shape desired. A skin of solid metal forms quickly at the mold-metal interface and shrinks from the mold wall; this reduces friction and allows the cast shape to settle continuously through the mold. Guides and pinch rolls regulate rate of settling of the cast shape and maintain proper alignment. As the casting protrudes downward from the bottom of the pinch rolls, it is cut to shape by a saw or oxyacetylene torch which travels briefly along with the casting. The cast piece is then laid to a horizontal position with a cradle; the casting may be rolled or otherwise worked directly without reheating if the plant is equipped with necessary facilities.

Certain features of the process are particularly critical: the mold must be as frictionless as possible and must remove heat from the metal at a carefully regulated rate relative to the casting speed and pouring temperature. Graphite molds are very satisfactory because they are self-lubricating: the metal does not wet them; also heat conductivity is good and if they are made massive enough water cooling is not necessary. Brass and copper molds of thin-walled tubing are used if water cooling is provided; a film of lubricating oil is kept between casting and mold wall, which reduces friction and acts as a heat-exchanging medium as the casting shrinks from the mold. In continuously casting steel ingot, the mold walls are $\frac{1}{4}$ in. thick if made of brass and $\frac{1}{16}$ or $\frac{1}{32}$ in. thick if made of steel; a stream of water is flowed at extremely high velocity over the outer mold wall to provide effective cooling. These values are not exact but are given for relative comparison.

Molten metal must be free from slag and poured with minimum turbulence to prevent surface roughness, which would cause undue mold friction and might interfere with balanced cooling and cause the casting to tear.

Heat must be extracted from the casting in such manner that directional solidification is obtained; otherwise, center-line shrinkage develops. This is done by graduated cooling, in which the rate of heat transfer increases down the length of the casting;

most heat is removed through the solidified metal, thereby allowing a cone-shaped pool of molten metal to exist continuously to
compensate for solidification shrinkage.

Advantages claimed for the process are (1) elimination of rough
forming and breakdown rolling operations, (2) 100 per cent yield,
(3) better surfaces than can be obtained with static ingots, (4)
improved quality due to reduction of segregation, (5) regulation
of grain size and structure by controlling cooling rates, and (6)
economy.

8.13   General considerations of non-ferrous ingots

Non-ferrous ingots are never cast with controlled gas content
as are rimming and semi-killed steels.   Dissolved gases are removed as completely as possible and solidification shrinkage is
compensated by adequate *directional solidification*.

Most non-ferrous alloys are subject to severe *microsegregation*.
Unless removed by heat treatment the segregation can result in
cracking during rolling and poor mechanical properties perpendicular to the rolling direction.   Also, the segregate can cause
ingot cracking during solidification.   The sizes and shapes of non-
ferrous ingots which can be successfully cast depends in large part
on the tendency of the cast alloy to exhibit cracking during solidification.   Non-ferrous ingots also exhibit *macrosegregation* which
may be *positive* or *inverse*.   In positive segregation, alloy elements
are rejected to the center of the ingot; in inverse segregation they
are drawn toward the surface.   Inverse segregation results directly from the "mushy" nature of solidification of non-ferrous
alloys, described in Chapter 5.   It may be so severe as to cause
low-melting eutectic liquid to *exude* from the casting surface.
These exudations must be removed by *scalping* before rolling.

Continuous or *semi-continuous* casting are the processes used
most widely today for casting aluminum and magnesium alloy
ingots.   They are also used extensively for casting copper-base
alloys.   Advantages of continuous as compared to static casting
are of both metallurgical and, sometimes, economic nature; they
have been described above.

8.14   Summary

Ingots have been discussed in some detail to provide a background of general information essential to a student's better under-

standing of the relationship of the cast ingot to the fabricated shape. Steel ingots are divided by internal structure into *rimming, semi-killed,* and *killed* types, which signify an ingot containing, respectively, primary and secondary blowholes and no pipe, fewer blowholes than a rimming ingot and a small shrinkage pipe, and no blowholes and a large pipe. The ingots may be *flat-sided* with square or rectangular cross section, or may be fluted or corrugated to increase cooling rate (surface area) and break up planes of weakness in crystal structure. Rimming and semi-killed ingots are usually made in massive, *big-end-down* metal molds, and killed ingots are made in *big-end-up* molds using a hot-top to supply molten metal to the solidifying ingot.

It was pointed out that blowholes are tolerable in ingots destined for drastic working because the holes are satisfactorily welded but that *blowholes cannot be tolerated in castings* because no plastic work is done upon them.

Detailed characteristics, advantages, disadvantages, and specific uses for all types of steel ingots were presented in this chapter; steel ingots were emphasized because their manufacture is relatively complex and their industrial importance very great.

Non-ferrous ingots are made in sand and metal molds, and ingot shape is of less consequence than with steel ingots. *Pitch* copper corresponds to a rimming ingot in that blowholes are purposely formed by oxidation (to remove sulfur) and by rabbling with a green wooden pole, usually a small sapling; all other copper ingots are dead set like a killed-steel ingot.

Continuous casting was discussed as a relative newcomer on the industrial scene. In this process bars, slabs, or cylinders are made by pouring metal continuously into one end of a water-cooled, bottomless die, thus forming a solid casting of desired shape which protrudes uninterruptedly from the other end of the die. The chief advantages of continuous casting are 100 per cent yield and the elimination of soaking pits and breakdown mills used for ingots.

8.15   Problems

1. List the requirements of a good ingot mold.
2. Illustrate with simple sketches the various types of ingots. What is unique for each type? Why are some ingot molds capped? Is it considered good practice? Explain.

3. Discuss the reactions involved in the production of rimmed steels; in killed steels. What are the principal factors controlling the rimming of steel? Why do we make semi-killed steel ingots?

4. Explain briefly why ingots should have smooth, defect-free surfaces. What defects are most detrimental?

5. What are the advantages and disadvantages of rimmed, semi-killed, and killed-steel ingots? List the uses of each type of ingot.

6. List the three types of volumetric shrinkage for steel. Describe a method for making shrink-free ingots.

7. Show by simple sketch the various crystal structures found in an alloy ingot.

8. Why do we find segregation in steel ingots?

9. What causes positive and negative segregation?

10. What is microsegregation? Where is it most prevalent, and can its effect be eliminated? If so, how?

11. List sources of non-metallic impurities in ingots. Discuss "hot-shortness" in steel ingots. How can it be minimized? What are the essential alloying agents in a free-machining steel?

12. How can indigenous and exogenous impurities be eliminated? What are the advantages of bottom pouring for ingots?

13. What precautions must be taken in casting aluminum and aluminum-base alloy ingots?

14. What is pitch copper? Why is copper cast in this condition?

15. Why are some ingots cast hollow? How can they be made?

16. Briefly describe the continuous casting process. List advantages realized by this unique process.

17. Suggest two physical causes for cracking of a steel ingot as it cools in the mold; also one reason for cracking after it leaves the mold.

18. Can you think of any way to minimize turbulence caused by top-pouring ingots, other than by bottom pouring? If so, explain.

19. Obtain some wax or stearin. After melting same and pouring into containers of various shapes and sizes, allow it to solidify and then section and study the nature and extent of the cavities resulting from solidification shrinkage.

20. Repeat Problem 19, using various means of chilling and insulating, and note all changes in the location and extent of shrinkage cavities.

21. Why do you think ingot molds have been made traditionally of cast iron? Why not use sand molds, for example? If sand molds were used, ingots could be made more like conventional castings with certain attendant advantages. Sketch your idea of an ideal process for ingot casting based on the utilization of sand molds, or mold made of similar granular refractory (i.e., other than metal molds).

Bring into consideration all the applicable knowledge you have gained from previous chapters on gating, risering, and casting processes.

22. It has often been said: "The most difficult casting to make sound (free of shrinkage defects) is a simple brick." Why should this be so? Is an ingot so very much different in principle?

23. For certain reasons of convenience steel mills would like to make killed steel ingots in the same manner as rimming steel ingots, i.e. big end down. This would be against the accepted principle of directional solidification. From heat-flow and heat-transfer considerations how do you think this desire of the industry could be realized and sound ingots of killed steel produced nevertheless?

8.16 Reference reading

1. Camp, J. M., and Francis, C. B., *The Making, Shaping and Treating of Steel*, U. S. Steel Co., Pittsburgh, 1951.

2. *The Solidification of Metals and Alloys*, A.I.M.M.E., New York, 1951.

3. Ruddle, R. W., *The Solidification of Castings*, Institute of Metals, London, 1957.

4. *The Principles of Basic Open Hearth Steel Making*, A.I.M.M.E., New York, 1950.

5. Atchison, L., and Kondic, V., *The Casting of Non-ferrous Ingot*, MacDonald & Evans, London, 1953.

CHAPTER 9

Casting Defects, Cleaning, Inspection, and Repair

PART I. CASTING DEFECTS, THEIR ORIGIN AND CONTROL

9.1 Introduction

Casting defects are usually not an accident; they occur because some step in the manufacturing process is not properly controlled. Melting, casting, and solidification of metal comprise many intricate operations, and perfect control is impossible; it is not surprising that castings are afflicted with a greater variety of defects than the products of other fabricating methods. To study defects rather than sound castings seems a negative approach to learning; however, it is important to producer, designer, consumer, and engineer that such defects as do occur in castings be recognized and their origin and control be properly understood.

9.2 Surface imperfections

Surfaces of castings are sometimes *rough* or *pebbly* because the molding sand is too coarse, or the pouring temperature of the metal too high. Sometimes *penetration* of metal between and around sand grains is a serious problem. In steel casting, a surface reaction occurs at the mold-metal interface which may cause roughness unless properly controlled. The reactions are the oxidation of the iron of the steel and subsequent combination of the oxide with silica to form fayalite:

$$2\,Fe + O_2 = 2FeO \tag{1}$$

$$2FeO + SiO_2 = 2FeO \cdot SiO_2 \quad \text{(fayalite)} \tag{2}$$

In magnesium casting, "burning" (oxidation) at the mold-metal interface causes a black, roughened casting surface. Burning is prevented by adding inhibiting agents such as sulfur, boric acid, or ammonium fluoborate to the molding sand.

Surface reactions sometimes cause subsurface porosity or *pinholes*. In aluminum alloys containing more than about 1 per cent magnesium, a reaction tends to occur between the magnesium of the alloy and the water vapor of the mold:

$$Mg + H_2O = MgO + H_2 \quad (gas)$$

Hydrogen-filled *pinholes* at the surface are the result. Some copper-base and other alloys are subject to similar defects unless proper precautions are observed. In steel castings subsurface pinholes can result from incomplete deoxidation of the molten metal.

Buckles and *scabs* are defects which, when present, usually appear in cope surfaces of castings. A buckle is a relatively extensive overlapping of metal; that is, a layer of metal is separated from the casting proper by a layer of sand which has sheared from the cope surface. Buckles and scabs are alike in appearance, the only difference being that scabs are relatively small. Usually these defects are caused by cope surfaces becoming preheated too rapidly by radiation from the rising metal. The non-uniform expansion of sand at and immediately behind the cope surface, caused by rapid preheating, develops stresses sufficient to shear layers of sand from the cope surface. These sheared layers then extend into the mold cavity and are surrounded by rising metal. The development of high gas pressure in a region of too low permeability can also produce excessive stresses in the cope and cause scabbing. Often cope surfaces preheated by metal actually *rain* sand down upon the rising liquid, causing sand spots. Tangent gates are used on large roll castings to swirl the metal rapidly and collect such sand particles at the center of the vortex, where they rise into feed heads. Buckles and scabs may be prevented by using sands with high hot-plasticity or low expansion characteristics, like chamotte, or by using an expansion buffer such as wood flour mixed with sand; the wood flour burns and leaves space for surface sand grains to expand without shearing from the mold. The problem may also be minimized by rapidly filling the mold.

Surface *laps, seams,* and *cold shots* result from pouring metal too cold. Usually these imperfections occur in castings with relatively light sections where two surfaces of flowing metal meet and do not fuse properly; they can be distinguished by clean appearance and lack of sand or other impurity to account for poor surface finish. In *cold shots,* small shotlike spheres of metal are almost completely distinct from the casting. These defects can usually be prevented by using higher pouring temperatures or by using streamlined sprues to give smoother flow, or a combination of these expedients.

Entrapped air, dross, and *slag inclusions* are usually found on the cope surfaces of castings. Entrapped air is recognized as small, internally oxidized holes just beneath the cope surface; it is found in aluminum and magnesium castings more frequently than in other cast metals. *Dross* and *slag* inclusions are oxides or other reaction products of the metal being cast (or the slag may be other non-metallics from the melting operation). These defects can be caused by improper control of (1) melting and pouring, (2) gating design, or (3) molding sand practice. To prevent slag or dross inclusions entering from the ladle, they should be skimmed before pouring, or teapot or bottom-pouring ladles should be used. Pouring basins should be employed so that any slag or dross entering from the ladle will rise and not pass into the runner system. The runner system itself should be designed to exert additional skimming action on the flowing metal. Streamlining the runner system will minimize any tendency to entrap air or form dross or slag inclusions during the mold-filling operation. Careful control of mold permeability and gas content reduce the danger of entrapping air (or mold gases) and minimize any tendency to form dross from mold-metal reactions.

Sand spots are caused by the metal washing particles from the runner system or mold walls. They appear as small irregularly shaped depressions in the finished casting and may be spaced randomly or gathered in clusters where the impurities have been collected at one or more vortices developed by the metal. Sand spots sink or float depending on the density of the metal cast; they may also become entrapped by growing dendrites along the sidewalls of the casting. Excess turbulence in the gating system, "spurting" of metal into the mold, and poor molding sand properties are causes of this defect. Methods of controlling foundry practices to eliminate the various defects described above are discussed in detail in Chapters 2, 7, 11, and 12.

9.3   Defects resulting from incomplete feeding

It is not widely appreciated that many of the more prevalent defects of castings result directly from solidification shrinkage; often the blame is laid erroneously elsewhere. Shrinkage of castings is noticeable in many forms; Fig. 9.1 shows typical defects

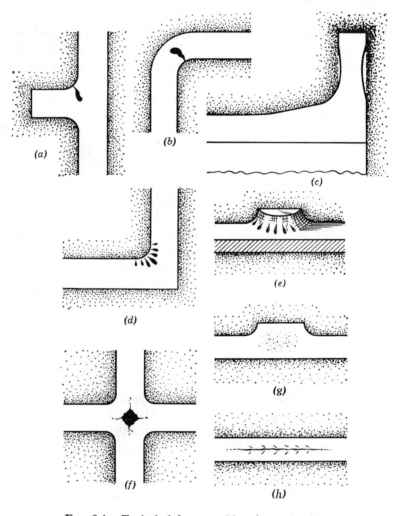

FIG. 9.1. Typical defects resulting from shrinkage.

due to shrinkage. The most prevalent type is localized cavities which appear at unfed hot spots in the casting. Whenever feeding is grossly inadequate, internal unsoundness is usually indicated by some external imperfection: either the wall punctures or deforms by *dishing* at the weakest point, or elongated *wormholes* appear at the base of the riser, at internal angles, or sometimes on cope surfaces. Sometimes shrinkage defects resemble collections of dross where cope surfaces are wrinkled and drawn inward. Often small voids which appear to be pinholes are really caused by imperfect feeding. If unusual surface imperfections, or depressed regions on cope and upper surfaces, are encountered, it is wise to consider shrinkage as a possible cause.

Gross shrinkage, center-line shrinkage, and *microporosity* are internal manifestations of solidification shrinkage. *Gross shrinkage* is a localized cavity most likely to occur in isolated sections called *hot spots:* i.e., the last regions of the cavity to solidify. *Center-line* shrinkage is a narrow, more or less continuous void sometimes found along the center line of castings with extensive platelike sections. The shrinkage may actually be in a continuous line or may appear as a line of fine *shrinkage chevrons* with the tips pointing away from the main direction of feeding. Center-line shrinkage is found only in alloys such as steel which freeze over a relatively narrow temperature range. Alloys that freeze over a wider temperature interval tend to exhibit *interdendritic* shrinkage (*microporosity*) when improperly fed. Microporosity also results from dissolved gas, and adequate *degassing* as well as directional solidification is necessary to produce a sound casting.

A less obvious defect resulting from imperfect feeding is *internal hot tearing.* *Internal hot tears* are radially disposed discontinuities inside castings, more commonly steel castings, which can be disclosed non-destructively only by radiography. The discontinuities resemble external hot tears, except that they are radial rather than roughly parallel (Fig. 9.2). The tears emanate from a low-density area, giving the radiograph an octopus-like appearance. It is usual to associate any form of hot tears with contraction stresses and to correct the fault by measures to give mold relief. *The cure for internal hot tears is improved feeding;* it has been found *very difficult to tear a sound casting internally* by stresses developed in normal casting procedure.

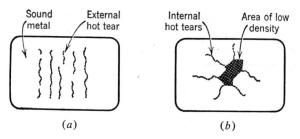

FIG. 9.2. Appearance of external and internal hot tears on radiographic film.

9.4 Gas porosity

Microporosity, center-line shrinkage and even gross porosity resembling gross shrinkage can be caused by gasses dissolved in the metal during melting and pouring, as well as by solidification shrinkage. The question "Is it gas or shrinkage?" is all too frequently asked. In some cases the distinction is relatively easy. Voids in a well-fed section must be caused by gas. Rounded voids with smooth walls are generally due to gas, while angular voids, perhaps with dendrite arms protruding into the voids, are usually caused by imperfect feeding. Often, as in the case of the fine microporosity seen occasionally in non-ferrous castings, the distinction is not simple, and both gas content and metal shrinkage must be regarded with suspicion when such microporosity is observed. Methods of eliminating gas porosity are described in Chapter 5.

9.5 External hot tears, cold cracks, and warpage

Unless castings can contract relatively freely in the mold as they cool from the temperature of solidification to room temperature, *external hot tears, cold cracks,* or *warpage* may occur. These defects result from (1) stresses developed in the casting by resistance of the mold (Figs. 9.3a and 9.3b), or (2) resistance of thinner parts of the casting to normal contraction upon them of heavier sections which cool more slowly (Figs. 9.3c and 9.3d). *External hot tears* are jagged, roughly parallel tears generally visible to the unaided eye but usually found by X-ray, magnetic powder, or Zyglo testing (discussed below); tears are often deep and are formed by surface rupture near the temperature at which the casting is completely solid.   The appearance of an external hot tear on a radiographic film is shown in Fig. 9.2a.

Cold cracks are somewhat similar to external hot tears; usually a cold crack is continuous, less jagged than hot tears, and forms in the cold casting from stresses imposed during cooling or heat treating.

Warpage is misalignment caused by stresses set up by conditions of cooling, improper design, or mold resistance; warpage does not involve any discontinuities such as is the case with cold cracks and hot tears.

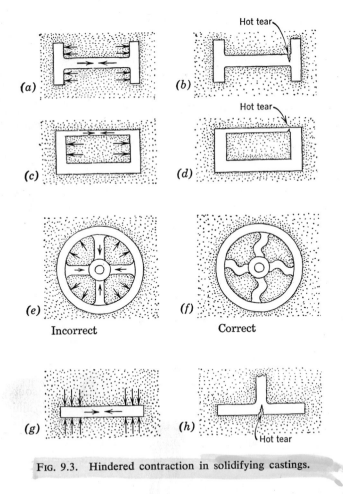

FIG. 9.3.   Hindered contraction in solidifying castings.

The obvious correction for mold resistance is (1) to increase *collapsibility* of molding sand by adjusting its composition, (2) to place chains or relieving blocks in the mold at critical regions so that they can be removed at the proper time to allow the casting to contract, or (3) to use coke or other filler which is less resistant than sand. Often stresses can be reduced by altering casting design, so that thin sections do not, by freezing prematurely, resist contraction of heavier sections. In circular castings, such as wheels with heavy spokes which tear longitudinally, tie bars and ribs are provided to strengthen the weaker members; sometimes spokes are curved rather than straight to reduce warping or tearing (Figs. 9.3e and f). Uniform sections of castings (Fig. 9.3g) do not tear readily, but areas where sections join (Fig. 9.3h) are potential sources of trouble. Tearing can also be prevented by using preheated molds, or by employing an alloy less susceptible to the defect.

9.6　Unfused chills and chaplets

Unfused chaplets and internal chills, as well as *blows* from chaplets and chills, are prevalent enough to make foundrymen wary of their use. Poor fusion is due to low pouring temperatures or the use of chaplets or chills so large they cool the metal before surface fusion can occur. Blows are due to moisture or other contamination such as oil, dust, or oxide accumulated on the chill or chaplet before or after it is placed in the mold. External chills sometimes cause surface blows by similar contamination. When chosen and used wisely, internal and external chills are of great advantage and give fully satisfactory results. Usually a clean, roughened surface, such as can be imparted by sandblasting, reduces blows and causes metal to lie quietly against external chills and to fuse with internal chills and chaplets.

9.7　Molding defects

Inadequate control of molding sand properties has been mentioned as a potential source of *sand spots*. Another source of this defect is the molding operation, from carelessness in finishing the mold or from use of damaged pattern equipment. Other defects resulting from improper molding include *oversize castings, mismatches,* and *fins*. Oversize castings result from rapping a pattern too hard when removing it from the mold, or from excessive allowance in the pattern for metal shrink. Also, sand rammed too soft can move, resulting in oversize castings from dilation; this is especially prevalent in gray iron castings (as described in Chapter 6). A *mismatch* is a slight displacement of the cope half of the casting with respect to the drag half; loose pins on the flask are the most common cause of this defect. *Fins* are a thin metal flash at the parting line of the mold or core halves; they are readily removed in the cleaning operation.

PART II.   CLEANING OF CASTINGS

9.8   Introduction

After castings are taken (*shaken* in shop vernacular) from the mold in which they are made, gates and risers must be removed, as does any refractory used for molds or cores that still might be sticking to the casting. This is the *cleaning* operation. Methods vary for different castings. Cleaning is often an expensive operation, and everything possible is done to reduce costs in this department. Use of *knock-off* cores, mentioned in Chapter 6, are an example of how money can be saved in the cleaning room; obviously any improvement in molding and casting practices, which results in better surface appearance (surface finish), also obviates cleaning problems.

9.9   Cleaning room practices

Table 9.1 is a concise summary of methods used for *cleaning* castings, from the time they are first shaken from the mold until they are finish-cleaned; also included are methods of preparing the castings for welding or repair. Study of these data will reveal a rather wide variety of methods used for cleaning castings. A knowledge of the properties of the different metals would immediately suggest reasons for the particular choice of tool. For example, flame cutting is not readily adaptable for removing gates and risers from copper-base castings because copper is not oxidizable enough to facilitate the action of the flame, and the thermal conductivity of the metal is so high that it is difficult, or impossible, to start a cut. Aluminum also has extremely high conductivity, and forms a refractory oxide skin which resists smooth

TABLE 9.1

SUMMARY OF METHODS FOR CLEANING CASTINGS

Cleaning Method	Aluminum and Magnesium				Brass, Bronze, and Copper			
	Removal of Refractories	Removal of Gates and Risers	Finish Cleaning	Preparation for Welding or Repair	Removal of Refractories	Removal of Gates and Risers	Finish Cleaning	Preparation for Welding or Repair
Wire brush	x				x			
Air-blast cleaning (sandblast, shot blast)	x	x			x			
Mechanical blast cleaning	x	x			x			
Water-blast cleaning	x	x			x			
Tumbling barrel								
Chipping hammer						x		x
Flogging								
Shearing		x						
Bandsaw		x		x		x		x
Hacksaw		x		x		x		x
Abrasive cut-off wheel								
Flame cutting								
Flame scarfing								
Flame gouging								
Grinding			x	x		x	x	x
Hand file			x	x		x	x	x
Rotary file			x	x		x	x	
Polishing			x			x	x	
Brushing			x			x	x	
Buffing			x			x	x	

burning; also aluminum can be cut so easily by saw that flame cutting would not be economical. Gray iron is not usually cut by flame because the carbon is so high that the metal is embrittled by the rapid cooling after cutting; also the high-carbon content makes cutting slow and inefficient. Since gray iron is quite brittle, gates and risers can usually be removed by "flogging"; a notch is left at the proper place on gate or riser and a

TABLE 9.1 (*Continued*)

SUMMARY OF METHODS FOR CLEANING CASTINGS

Gray Cast Iron				Malleable Cast Iron				Ductile Cast Iron				Steel			
Removal of Refractories	Removal of Gates and Risers	Finish Cleaning	Preparation for Welding or Repair	Removal of Refractories	Removal of Gates and Risers	Finish Cleaning	Preparation for Welding or Repair	Removal of Refractories	Removal of Gates and Risers	Finish Cleaning	Preparation for Welding or Repair	Removal of Refractories	Removal of Gates and Risers	Finish Cleaning	Preparation for Welding or Repair
x				x				x				x			
x				x				x				x			
x				x				x				x			
x				x				x				x			
x	x			x	x			x				x			
		x							x	x			x	x	
	x				x										
	x		x		x		x		x		x		x		x
													x		x
														x	x
														x	x
		x	x			x	x			x	x			x	x
		x				x				x				x	

hammer blow usually suffices for removal; the stub is then ground flush to the casting surface with a grinding wheel.

The following paragraphs discuss the cleaning practices that are most widely used in foundries today. Figure 9.4 illustrates the cleaning operations that would be performed on a typical small aluminum casting, and Fig. 9.5 illustrates steps in cleaning a large steel casting.

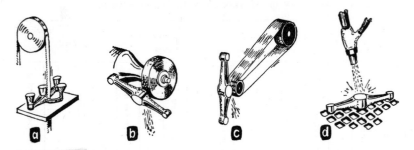

FIG. 9.4.   Steps in cleaning a small aluminum casting (lever arm).   (a) Gates and risers removed by bandsaw. (b) Rough grinding (snagging) with coarse abrasive wheel. (c) Finish grinding (polishing) with fine abrasive belt. (d) Sand blasting.

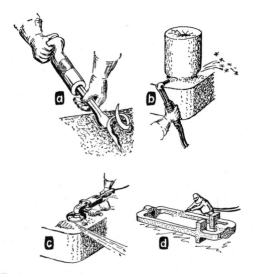

FIG. 9.5.   Steps in cleaning a large steel casting (press frame). (a) Fins and burned-on sand. (b) Gates and risers removed by flame cutting. (c) Riser pads and gate stubs finished with abrasive disk. (d) Casting shot-blasted.

9.10 Removing the mold material

After shakeout, any adhering sand or refractory is removed with a wire brush, by jolting the casting against a firm object, or by *blast cleaning.* Three types of blast cleaning are in common use in foundries; they are *air, water,* and *mechanical* blasting units. In air blasting, a high-pressure stream of air propels sand (*sand-blasting*) or shot (*shot blasting*) against the casting surface. In water-blast cleaning, water is used instead of air, to propel sand. In the mechanical units, rapidly rotating paddles mechanically hurl shot or sand on the surface to be cleaned.

Blast-cleaning units are available in a variety of different types for special needs. In small cabinet-type sandblasters, the operator reaches into the cabinet with rubber gloves and moves small parts to be cleaned back and forth under the blast. In larger sand- and shot-blast units, a special ventilated room is set aside; the operator (protected by dust mask and special clothing) directs the blast toward the casting from a flexible hose. In modern blasting systems for large castings, full control of the blast is possible from outside the room. Modern automatic units for smaller castings rotate or tumble the castings under the grit blast. Figure 9.6 illustrates one such unit where abrasive is mechanically blasted onto castings which are being tumbled on an "apron conveyor."

The *tumbling barrel,* another method for cleaning adhering sand from castings, was once much used for gray iron and still finds

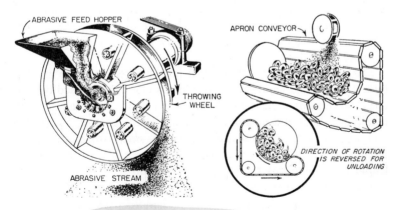

ABRASIVE FEED HOPPER

APRON CONVEYOR

THROWING WHEEL

DIRECTION OF ROTATION IS REVERSED FOR UNLOADING

ABRASIVE STREAM

Fig. 9.6. Mechanical blast cleaning (Wheelabrator system).

some application. A rotating iron drum is filled with castings mixed with small, star-shaped, hard iron pieces called *mill stars*. Sometimes pieces of graphite electrode are also added. As the barrel rotates, sand is rubbed or shaken from small castings, gates are broken, and all surfaces are liberally burnished; the purpose of the bits of electrode is to give a good surface appearance and possibly to make small surface imperfections appear less objectionable. Batch operation of an old-style tumbling barrel was a slow, cumbersome, dusty process; today such barrels are filled and emptied continuously by belt, or chain conveyors—they are often kept wet with a steady flow of water to reduce dust.

9.11   Removing gates and risers

After removal of adhering sand or refractory by one or another of the preceding techniques *gates and risers* are removed from the casting. As shown in Table 9.1, a wide variety of methods are available for accomplishing this, including chipping, flogging, shearing, sawing, abrasive wheel cutting, and flame cutting.

Chipping hammers are air-driven hammers with a chisel as the working tool. They are very versatile, but are subject to high human equation in efficiency of operation. The sharpness of the cutting bit, the way it is ground to shape, the material of which it is made, and in particular the way it is held against the work by the operator affects cutting efficiency greatly; keeping the cutting tip wet with oil can mean as much as a 40 per cent improvement in operation. Chipping hammers are useful for low-, and medium-carbon steel and copper-base castings where hardness and ductility are suitable to the process. For more brittle materials, gates and risers are more easily removed by *flogging* (striking with a hammer or sledge).

Shears are heavy, matching steel jaws worked past one another by a flywheel and cam assembly. Shears can only be used on materials softer than the shear blades; this limits their use to malleable iron, soft and medium hard steel, brass, bronze, aluminum, and magnesium. Shears are limited to small work, but when adaptable are fast and economical. They are often used in malleable iron foundries for removing the stub left when gates are broken from the casting.

Bandsaws are being used more and more extensively as new knowledge is gained about blade design, thus adding to the range of metals that can be cut; they are used most extensively today for cutting non-ferrous metals, but saws are now available for cutting mild and low-alloy steels. It is very important to match the type of blade, particularly the number of cutting teeth per inch, to the metal being cut, and to keep the blade well oiled if required. Bandsaws are fast and economical whenever the work is small enough to be held by hand. For larger work automatic hacksaws are used; the above observations in operating bandsaws apply equally to hacksaws.

Abrasive cut-off wheels are used whenever metals are too hard to be cut by saw, and where flame cutting or chipping is not feasible. Abrasive cutting is generally more expensive than other methods, although new developments in wheels may bring abrasive cutting into closer economic competition with other methods of metal removal, particularly if careful account is kept of all costs. The smooth cut left by the wheel needs no further attention if the cut is made flush with a flat surface of the casting. For obvious reasons the cut-off wheel is limited in versatility because, more often than not, cuts must be made on other than flat surfaces.

Flame cutting is an extremely versatile method for removing metal. Flame cutting was once relegated to those metals readily oxidizable by the flame, and to plain, lancelike cuts. Today, even stainless steel, which is difficult to oxidize, can be cut by torch. To do this, iron (in powder or rod form) is fed into the flame at the tip of the torch to provide an oxidizable medium; this has proved beneficial for cleaning steel castings where the mold was penetrated by molten metal and layers of sand locked (burned on) to the surface of the casting. *Flame gouging* and *flame scarfing,* for removing excess metal, are adaptations of flame cutting that should find increasing use in foundries; tips are provided for regulating gas flow at higher volume and lower velocity than are needed for cutting, and the tips are curved to accommodate the work. It is possible to *burn* (melt) metal to leave small, deep, or shallow holes, or metal can literally be *washed* in large volume from the surface. Flame scarfing and gouging are much used for removing surface defects from ingots, and are being used more and more for removing riser pads, cleaning away penetrated sand, and for preparing castings for welding. Flame cutting is not commonly used on small castings, or on castings that would be damaged by the heating effect of the torch.

9.12   Finish cleaning

After removal of gates and risers, castings are given one or more *finishing operations*. These are performed to (1) smooth the gate and riser areas of the casting, (2) remove any excess metal remaining on the casting, and (3) improve surface finish or appearance.

Preliminary metal removal is obtained by *grinding*. An abrasive grit of about 25 to 80 mesh is bonded into the shape of a wheel (*grinding wheel*) or onto a belt of paper or canvas (*abrasive belt*). Figure 9.4 illustrates cleaning by grinding. Wheels and belts are available in a variety of different shapes and sizes, and of different grits, grit sizes, and binders; it is important to choose the right combination for a given application. Grinding machines are either portable or stationary; the portable machines are usually used only for large castings. In the softer metals *rotary files* are sometimes used instead of grinding. A rotary file is essentially a hardened steel grinding wheel. In steel and cast iron, *chipping* is employed as an adjunct to grinding to remove fins, pads, and other protuberances.

If a smooth surface finish is desired, castings may be *polished, brushed,* or *buffed* after grinding. *Polishing* is usually performed with an abrasive belt machine; it is similar to abrasive belt grinding except that a finer grit is used (80 to 400 mesh). *Brushing* employs rotary wire or fiber brushes that remove burrs and grinding marks. *Buffing* is used to obtain an exceptionally high luster on cast surfaces. It employs wheels of "buff" (usually disks of muslin sewn together); a very fine abrasive mixed with a grease binder is rubbed on the face of the wheel.

Unless castings are brushed or buffed, they are usually given a final *blast cleaning* (or *tumbling*) before shipping. These operations remove dirt and grease, and smooth any remaining grinding or polishing marks. In blasting, care must be taken to use a sand or shot grit which is sufficiently fine; otherwise the operation may actually increase the roughness of the as-cast surface. For castings with a very fine as-cast finish, as in plaster or investment molding, it is sometimes advisable to use grits such as rice hulls, ground corncobs, or ground walnut shells in the blasting operation, instead of the sand or shot normally employed.

9.13 Introduction

As soon as castings are clean enough, they are inspected by eye, gross defects are repaired if necessary, and the castings are given further inspection. Nearly all inspection methods applicable to metals in general can be applied to check and control quality of castings. Radiographic and other critical methods are used to increasing extent, particularly if service performance for the castings is severe; for example, for cast steel valves used in steam systems at 1200°F (650°C) and 1400 psi pressure. For such applications castings must be dependably sound. Very rigid standards (such as *Radiographic Standards for Steel Castings,* U. S. Bureau of Ships, Navy Department, Washington, D.C.) have been developed for radiographic inspection. Recently similar standards have been developed for aluminum and magnesium alloy castings,[1,2] and it is probable that precise standards will eventually appear for fluorescent and magnetic powder inspection. Originally radiographic inspection of castings was desired only by the consumer as insurance of casting integrity. Foundrymen have now found that such exacting tests can aid greatly in their efforts to improve their product; for quality work, most large foundries now utilize every critical test available to them.

The use of statistical quality-control methods for inspecting castings is a promising new development for foundries. By statistical methods of sampling only a calculable number of production castings are checked critically. This saves time and money and often proves to be as adequate as checking 100 per cent of the production.

9.14 Specific methods for inspection

Some common inspection methods are the following: *Non - Destructive*

1. *Visual examination.* All too often substantial time and money are spent cleaning castings which are visibly defective. Often a visual inspection station at the shakeout platform is warranted.

2. *Magnetic* and *fluorescent powder inspection.* In the former, magnetic powder is either dusted over the surface of a magnetized casting or flowed over it in an oil suspension. Powder particles

collect at any crack or discontinuity in or near the surface; opposing free surfaces of the defect act as north and south poles. The whole casting may be magnetized, if enough current is available, or *prods* may be used to inspect small areas of large castings. By changing the prods systematically the entire surface of large castings can be surveyed.

Castings of non-magnetic materials may be immersed in a warm suspension of fluorescent powder in penetrating oil; the suspension penetrates even minute cracks and pores. Excess suspension is washed away in warm water, and the casting is dusted with a drying powder; this draws some of the suspension from the cracks to the surface, where it fluoresces and is readily visible under ultraviolet light. *Die-penetrant inspection* is a variation of fluorescent powder inspection wherein a colored die is suspended in penetrating oil, rather than fluorescent powder. The use of ultraviolet light is thus obviated, but defective areas are somewhat less readily seen.

Magnetic powder inspection is useful for detecting discontinuities at and slightly below the surface. It can be used only on ferrous materials. Fluorescent powder inspection can be used on all materials, but detects only *surface* defects. Both inspection methods detect fine surface cracks and flaws more readily than radiographic examination; they are also considerably cheaper than radiography.

3. *Radiography* can be utilized for inspecting castings of any metal; the techniques are not materially different from those in medical investigations. Radiant energy from an *X-ray tube,* capsule of *radium sulfate* (radium or gamma-ray radiography), or *radioactive cobalt* is passed through the casting, or section of the casting, and recorded on a film held against the opposite surface. Defects in the form of cracks or voids are recorded as blackened areas on the film, since the radiant energy moves more easily through the less dense regions. To interpret radiographic films of castings correctly requires considerable skill and experience, as in medicine. Some defects readily disclosed by radiographic inspection are sand spots, cracks, internal and external hot tears, unfused chills and chaplets, shrinkage, and subsurface gas or pinhole porosity—in short, any discontinuity or collection of discontinuities exceeding about 2 to 3 per cent of the casting thickness.

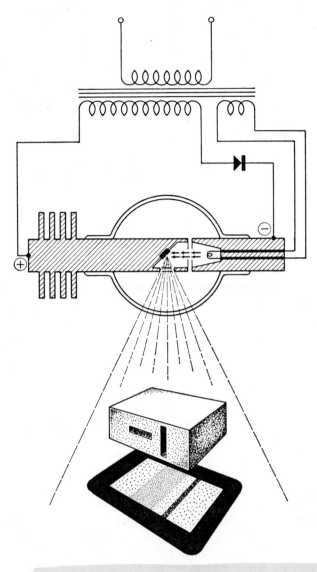

FIG. 9.7. X-ray inspection. Top, high voltage transformer and X-ray tube; center, casting with internal voids (black); bottom, appearance of X-ray film. Blackening of film is proportional to thickness of the internal defect.

Figure 9.7 illustrates the appearance of an X-ray film of a casting with internal voids.

Inherent characteristics of radium and X-ray radiography may make one or the other desirable for a particular application. X-ray radiography as compared with radium radiography enjoys *shorter exposure times* (usually minutes rather than hours as required with gamma rays), and *better definition* and consequently *clearer images.* X-ray equipment is relatively very high in *first cost* and *maintenance,* has limited *maneuverability,* and requires that castings be brought to the machine. Radium is particularly applicable when castings are extremely large or inspection must be made at *the site;* in addition, many castings can be *radiographed simultaneously* by placing them in a circle around the capsule, and overnight exposures may be taken *without continuous supervision.* Radioactive cobalt can be procured from government sources and has characteristics similar to radium. Radium pills can be rented for unlimited times at low cost.

4. *The ultrasonic reflectoscope* is sometimes used for finding subsurface defects such as cracks and shrinkage cavities in large castings. A flat surface is prepared on the casting against which a quartz crystal is pressed; a thin film of oil insures good contact. Ultrasonic vibrations are sent through the crystal, and their reflections from any free surface are picked up and recorded on an oscilloscope. The position of the defect within the casting can be found accurately by a calibrated scaling system on the oscilloscope.

5. *Ring tests or vibration tests* reflect the soundness of a piece of metal by the quality of the "ring" it produces. It is an easily conducted but rather crude test when the "ring quality" is determined by the human ear. Recently inspection equipment has been developed to "listen" to the sounds electronically and compare the tone with that of a casting known to be sound.

6. *Dimensional control* is nearly always required; sometimes it is an extremely critical requirement. Initial castings from a new pattern are usually carefully measured in a *layout room* to be sure the casting conforms to blueprint specifications. After castings are found to be consistently within tolerances, spot checks suffice for production control. Often elaborate gages are specially constructed to permit a rapid check of particular casting dimensions during a production run.

7. Smooth *surface finish* is always a desirable characteristic and is sometimes a rigid requirement. It is described by a "number" which is an average of the heights of the peaks (and depths of the valleys) on a casting surface. The number may be a simple "arithmetical" average or it may be a "root mean square" average; the differences between these two averages are fully described elsewhere.[3] In practice, the two types of averages are very nearly the same (within 10 or 15 per cent).

Casting surface finishes vary from about 90 microinches for smooth plaster molded castings to over 300 microinches for heavy castings made in coarse sand. The degree of surface roughness on as-cast surfaces is determined by *visual comparison* with cast standard, or by use of a direct-reading *surface roughness indicator* (Profilometer). In the later case a stylus is drawn carefully over the surface; an electronic pickup translates the signal from the stylus into an average surface roughness reading. Visual comparison (with a standard) is an adequate quality-control check for most sand castings but very smooth-surfaced sand castings and plaster or investment castings require the use of an automatic indicator to accurately gage surface quality.

8. The inspection methods described above are non-destructive tests, in the sense that castings need not be sacrificed to the test. *Transverse radiography* is a test in which one or more thin slabs (usually ⅛ to ½ in. thick) are cut from the casting at suitable points and radiographs are taken across their width. In this way, a defect which might constitute much less than the minimum detectable in the entire cross section is integrated to its maximum effect. Only *pilot castings,* used to develop the casting method, and an occasional check casting from production would be sectioned and explored so critically. This test is particularly useful for checking center-line and interdendritic shrinkage and for detecting finely distributed gas porosity which does not show up readily by non-destructive methods.

Other tests which fall more in the category of acceptability standards, but which are often considered inspection tests, are listed below; these will be referred to again in Chapter 13 as aids for quality control.

1. *Chemical analysis* to determine whether composition is within allowable limits.

2. *Metallographic appraisal* as a check on grain size, non-metallic impurities, submicroscopic pinholes, type and distribution of phases present in the cast structure, and response to heat treatment (if heat treating is required).

3. Mechanical testing methods, including (*a*) *tensile test* to determine ultimate tensile strength, yield strength, elongation, and reduction of area; (*b*) *bend, notch bend, impact,* and *transverse tests* to appraise ductility and resistance to shock; (*c*) *fracture* of the rough casting to check apparent soundness; (*d*) *pressure testing* of as-cast or machined unit to pressures in excess of service requirements.

4. *Hardness* is often a measure of acceptability of castings; hardness is determined by measuring resistance of a metal to deformation by penetration with a steel ball, or indentor. Hardness reflects ductility and tensile strength of sound metal with reasonable accuracy. Sometimes the only inspection method used, besides visual standards, is a hardness measurement. Standard types of hardness testers are available; of the many types the Rockwell, Brinell, and Scleroscope units are most useful for checking castings.

PART IV.   REPAIR OF CASTINGS

9.15   Introduction

Having studied the many steps in making a casting, the nature and extent of defects in castings, it is obvious that repair is an essential phase of foundry operations. Repairs may be necessary because the casting is torn in the mold or broken in handling, because of minor or extensive surface imperfections, or because of one or another manifestation of shrinkage. It is not uncommon for repairs to more than double the cost of the casting and yet for the process to be economical. For example, in one actual case all molding, melting, casting, and cleaning operations on a particular large casting was estimated to be $4000, and repairs $5000, making a total cost of $9000; yet casting plus repairs compared favorably with an estimated cost of $22,000 to make the engineering part by fabrication methods other than casting.

Often too much reliance is put on repairing castings and too little attention given to making the casting better initially. For example, castings are sometimes made with full knowledge that they will be porous, and will require impregnation of the entire part with resin, Bakelite, or solder. Fortunately more and more foundrymen are taking the more positive course of improving their product and reducing their reliance on repairing.

9.16   Specific methods of repair

Some specific methods of repair are the following:

1. *Arc and gas welding.* In repair by these processes the area to be welded is first cleaned by grinding, filing, chipping, gouging, or scarfing as listed in Table 9.1. Heat from an electric arc or gas torch is then used to melt the base metal around the area requiring repair; at the same time "filler" metal is added from a welding rod. If the filler metal is of the same composition as the base metal, and particularly if the casting is heat-treated after welding, the weld area can usually be made fully as strong and ductile as the casting itself. Many specific types of arc and gas welding processes are used for repair; they are described in detail in Chapter 14.

2. *Exothermic welding.* Large shrinkage cavities, cracks, and other forms of gross porosity are sometimes repaired with exothermic methods. This technique is similar to that described for exothermic casting in Chapter 5. The area to be repaired is cleaned as for standard welding, and a dam of sand or other refractory built around it. For repairing steel castings, a mixture of iron-oxide, aluminum, and alloy powders (blended to give proper composition in the filler metal) is either charged into a separate melting chamber, or used to fill a large mold above the defective area. The mixture is ignited "in situ" or reacted in the separate vessel and the molten metal formed exothermally (without addition of external heat) is poured into the mold. Either method is used satisfactorily. Metal is formed by the exothermic reaction

$$Fe_2O_3 + 2Al = Al_2O_3 + 2Fe$$

at about 4500°F (2500°C), and this metal settles readily from the Al_2O_3 slag. The porous area is melted by the high heat and a sound weld is formed. The weld deposit can be made to the chemistry desired, and is not limited to ferrous compositions; by using other exothermic reactants, non-ferrous alloys can also be repaired.

3. *Flow welding,* or "burning in." Many times casting defects such as localized shrinkage cavities are repaired by remelting and recasting "in situ," using metal from the ladle. The casting is embedded in sand, or a dam of sand molded around the defective area; molten metal from the ladle is then poured directly into the cavity, or flowed over it continuously, until melting takes place. Pouring is then stopped, the casting cooled, and the excess metal ground or machined to dimensions. Many pounds of metal are flowed over the defect for each pound left in the weld, and so this usually develops as an expensive source of heat. The practice was once widely employed in foundries but fell into disuse when other welding methods were perfected. Today it is again finding favor for certain applications, particularly for repair of high-quality non-ferrous castings. Close control of composition of the weld deposit is possible with this method, and a sound, dense weld results. Danger of weld cracking is minimized, and by careful molding of the flow channels it is possible to build intricate "weld" deposits onto defective castings.

4. *Braze welding.* Braze welding involves the use of a filler material which melts above about 800°F (430°C) but below the temperature of the base metal. In this process the base material is not melted and the bond between weld and casting comes from a small amount of alloying (by diffusion). Since the base material is not melted, the defective area must be thoroughly cleaned, and a *flux* must be used to remove surface oxides before deposit of the filler material. Braze welding is employed to repair cast parts which would tend to crack or distort if other welding processes were employed. One example is the use of yellow brass to repair defects in gray iron.

5. *Soldering.* Soldering is similar to brazing except that it employs filler materials which melt below 800°F (430°C). Most solders are alloys of tin and lead with small amounts of other added elements. In repair work, soldering is used primarily to fill small surface imperfections when high mechanical properties are not a prerequisite.

One unique repair by soldering is filling porous areas of copper-base castings, particularly the tin bronzes. A good example is bronze bearings; these are sometimes blessed with pinholes or interdendritic porosity, and leak liberally under pressure testing after machining. To repair such castings they may be *impregnated* with Bakelite or resin (discussed below), or *wiped* with soft solder.

This technique has been applied, with satisfactory results, to castings weighing several tons. Machined surfaces are first cleaned with hydrochloric acid or other soldering flux. The casting is then heated to above the melting temperature of the solder (but below that of the casting), and solder is wiped over the machined surface with an asbestos rag; an alternative method is to dip the portion of the casting to be soldered into the solder bath. In either case, the molten metal flows into the voids; the casting is then cooled and the surfaces are scraped or remachined to prepare them for service.

6. *Impregnation.* If porous areas in castings are fine, such as interdendritic voids or pinholes, and are too extensive for welding or brazing to be economical or effective, impregnation with Bakelite or resin is indicated (unless "wiping" with solder is used). Steel castings have been successfully impregnated with Bakelite, and light metal castings are extensively repaired by resin impregnation.

In either process the casting is thoroughly cleaned, and usually machined in critical areas to "open up" porosity as favorably as possible (many iron and other castings, however, are treated unmachined). The casting is heated for Bakelite treatment, but this is not essential for resin impregnation. It is immersed, under vacuum, in a tank of liquid Bakelite or resin, after which it is dried in air and heat-treated as required to "set" the impregnant permanently. Several companies have been formed to impregnate castings on an industrial scale. The manufacture of castings so grossly porous as to need such general repair is deeply deplored, but the process is important enough that government specifications have been drafted for the practice.[4]

7. *Plastic fillers.* It has long been customary to fill porous areas, cracks, and other surface imperfections in iron castings with a material known to the trade by names such as "Smooth-on." This is a pasty mixture of iron filings in a hardening agent, which "sets" in air to very nearly the hardness and color of the casting. It is difficult to notice repairs of this sort on the surface of iron castings, particularly if the part is later ground or machined and cleaned by tumbling or shot blasting.

9.17   Repair of ferrous castings

Steel castings are almost universally repaired by welding. In recent years welding repair has been extended more and more to iron and high-alloy castings, and it has become increasingly easy to weld hardenable grades of these metals; this has been brought about by improved knowledge of welding methods, and wider appreciation of the effects of phase transformations and gases on weld deposits and on parent metal. Metallurgical difficulties encountered in welding steel castings are discussed in Chapter 14.

Defects are prepared for welding by grinding or chipping or by flame gouging or flame scarfing. Sometimes small surface defects noticeable after blast cleaning are "touched up" with a small weld deposit without special preparation. It is always necessary to remove all defective metal before attempting extensive repair by welding. For example, a hot tear, only partially removed, will become aggravated by the weld deposit; also gases contained in the crack or porous area will expand upon heating and cause imperfect welds. It is often difficult, as in grinding or chipping, to be sure that all of the defect is removed, because the metal is often flowed or "burred" over the region, hiding imperfections beneath. Flame gouging or scarfing is excellent in this respect because the defect shows as a darker region than the surrounding hot metal, because of differences in heat transfer through voids and sound metal.

Electric-arc welding is almost universally used for repairing steel castings. *Shielded arc welding* (arc welding with coated electrodes) is the most widely used type but *inert-gas metal-arc welding* (bare metal consumable electrode with inert gas shield) is gaining favor for repairing large defects. Specialized procedures and equipment are often necessary for the higher-alloy steels; some of these are described in Chapter 14, but the welding handbooks [5, 6] should be consulted before attempting repair of the alloys.

Gray cast iron can be repaired by most of the standard fusion welding processes, but careful preheating and slow cooling are necessary to prevent (1) weld cracking and (2) transformation of the structure to hard, brittle martensite. Gas welding is usually found to be preferable to minimize these effects. For applications where corrosion or thermal stresses are not a problem, or where a color match is not necessary, *braze welding* with yellow brass is a desirable repair method. The strength of the joint is fully comparable to that obtained with a cast iron filler metal.

Malleable iron castings are seldom welded in the annealed or heat-treated condition because of the difficulty of achieving a ductile weld. In the as-cast state (white iron), they can be readily repaired using white cast iron welding rods. Gas welding is usually employed. The weld solidifies as white iron and malleableizes equally as well as the base metal; after cleaning, the repaired castings are satisfactory in all respects.

Repair methods other than those mentioned above are used to a limited extent in salvaging iron and steel castings. These methods have been described in Section 9.15.

9.18 Repair of non-ferrous castings

Aluminum-, magnesium-, and copper-base alloy castings can be readily repaired by welding. Gas welding, metal-arc, carbon-arc, or inert-gas metal-arc welding processes are used for aluminum- and copper-base alloys. Only inert-gas-shielded arc welding processes are recommended for magnesium alloy castings. Chapter 14 describes these in detail.

Aluminum and magnesium castings are not repaired by braze welding, but surface defects are sometimes filled with solder unless this is prohibited by specification. Brasses and bronzes can be repaired by braze welding or soldering. Impregnation and flow welding are used for repairing all types of non-ferrous castings.

9.19 Summary

In this chapter the casting process is considered in the interval after castings are shaken from the mold and before they are heat-treated, machined, or shipped to the customer. In the first section defects likely to occur in cast metals are considered. It is desirable for the designer, consumer, and engineer to appreciate why defects occur in cast metals, how they are cured, and, more important, how they can be prevented.

Since the casting process is not subject to 100 per cent perfect control, *inspection* is a vital step to prevent defective parts from reaching service. Casting inspection comprises nearly all the methods normally employed for investigating metals:

1. Visual examination.
2. Magnetic and fluorescent powder inspection.
3. Radiography.
4. Ultrasonic reflectoscope survey.
5. Ring tests, vibration tests.
6. Dimensional and surface finish examination.

7. Transverse radiography.

8. Other methods, including chemical analysis, metallographic check, hardness testing, and mechanical testing methods.

The types and amount of inspection performed on a casting depend in part on the control exercised in the manufacture of the casting, and also on the end use of the cast part.

After castings are taken (*shaken* in shop vernacular) from the mold in which they are made, they are *cleaned* by (1) removing any sand or refractory still sticking to the casting, (2) removing gates and risers, and (3) finishing the casting surface so as to remove any remaining undesired metal and improve the casting appearance. Also, in the cleaning operation, defects are removed preparatory to repair by welding, brazing, or soldering. Table 9.1 is a concise summary of the methods used for cleaning castings of different metals.

Repair of castings is an essential phase of foundry operations; often repairs can more than double the cost of a casting, and yet the process will still be economical. By proper care in procedures, repaired castings (of most metals) can be made fully as strong and ductile as castings that do not require repair. In spite of the obvious advantages of casting repair, care must be exercised to assure that repair techniques do not become a substitute for good foundry practices.

Specific methods of repair described in this chapter are:

1. Arc and gas welding.
2. Exothermic welding.
3. Flow welding.
4. Braze welding.
5. Soldering.
6. Impregnation.
7. Use of plastic fillers.

9.20 Problems

1. Discuss briefly the mechanism of metal penetration. How can it be minimized?

2. What is surface drossing? What ways can you think of to eliminate or alleviate the condition? Are all your methods practical (that is, can they be done economically in practice)?

3. What causes "raining" of sand during pouring of large cast-

ings? Hold a flat cake of core sand, or rammed molding sand, over a blow torch and note the result.

4. Briefly discuss advantages of pouring metal as cold as possible.

5. What are the causes of external hot tears, cold cracks, and warpage? How would you prevent their formation?

6. Discuss internal hot-tear formation. How would you prevent hot-tear formation in a flat, rectangular plate casting?

7. What causes surface shrinkage at internal angles in castings? How can it be eliminated? Pour molten wax or stearin in a metal box, and cover with a metal plate in contact with the upper surface. Then repeat, using a small piece of "felt" attached at a convenient location. Note the position of the surface shrinkage in each case.

8. Discuss briefly center-line shrinkage. What is the marked difference between center-line and gross shrinkage?

9. What is interdendritic shrinkage, and why does it occur? Why does it cause the non-ferrous foundryman so much trouble?

10. What precautions should be taken when using chills and chaplets?

11. How would you gate a gray iron casting to achieve greatest economy in the cleaning room? How would you gate an aluminum casting?

12. List the cleaning operations you would perform on a bronze casting to be used for decorative purposes.

13. List the cleaning operations you would perform on a large steel casting to be used for structural purposes.

14. List four rules of risering which should be followed to minimize cleaning costs.

15. List three types of casting inspection (other than mechanical methods), and briefly discuss why they are used.

16. In what type of work would the following test procedures be followed: (1) X-ray radiography; (2) gamma-ray radiography; (3) ultrasonic testing; (4) fluorescent oils and powders; (5) magnetic powder testing; and (6) pressure testing? Give simple reasons for your answers.

17. You are producing high-quality aluminum castings for aircraft use. It is important that each one be completely sound and free from surface defects. What inspection tests would you perform, and in what order would you perform these tests?

18. Why is repair welding more frequently conducted on large steel castings than on other types of castings?

19. What are the major advantages of repair by exothermic welding?

20. List the steps you would follow in repairing a shrinkage cavity in a low-carbon steel casting.

21. What are the advantages of repair by flow welding? the disadvantages?

22. Why are gray iron castings sometimes repaired by brazing rather than by arc welding?

23. Obtain bar magnets of suitable types and shapes. Place a sheet of white paper over the magnet tips when in contact and nearly in contact; then sprinkle iron filings over the paper. The principle of magnetic particle testing will be evident.

24. Scrap selection and sorting is an important and valuable process. Discuss the simplest possible means you would use to sort the following miscellaneous metals when all are mixed together.

(a) white iron.   (f) lead.

(b) pure iron.   (g) aluminum.

(c) pure lead.   (h) railroad rails.

(d) stove plate scrap.   (i) austenitic stainless steel.

(e) ductile iron.   (j) ferritic stainless steel.

25. Gates and risers can be cut readily and without damage to the casting (with an oxyacetylene torch) from plain low-carbon steel castings but not from high-carbon steel castings. Why? What added feature is needed to make this process applicable for stainless steel?

9.21   Reference reading

1. *Aeronautical Technical Inspection Manual,* Navaer 00-15PC-504, Naval Aviation Supply Depot, Philadelphia.

2. *Tentative Reference Radiographs for Inspection of Aluminum and Magnesium Casting,* E98-53T, A.S.T.M.

3. *Surface Roughness, Waviness, and Lay,* A.S.A. Standard B46.1-1955, Technical Bulletin, Micrometrical Mfg., Ann Arbor, Mich.

4. *Impregnation of Porosity in Castings, Process for,* MIL-STD-276, General Services Administration, New York.

5. *Procedure Handbook of Arc Welding Design and Practice,* Lincoln Electric Co., Cleveland, 1950.

6. *Welding Handbook,* A.W.S., New York, 1949.

7. Briggs, C. W., *The Metallurgy of Steel Castings,* McGraw-Hill, New York, 1946.

8. *Analysis of Casting Defects,* A.F.A., Chicago, 1947.

9. Rossi, B. E., *Welding Engineering,* McGraw-Hill, New York, 1954.

10. Newell, W. C., *The Casting of Steel,* Pergamon, New York, 1955.

CHAPTER 10

Pattern Construction and Casting Design

10.1 Introduction to pattern construction

"As the pattern goes, so goes the casting." This phrase is as important as it is simple, and is a truth as well known to foundrymen as it is widely ignored. The quality of a casting is influenced by the material from which the pattern is made, and by how well the pattern maker and the engineer who laid out the gating and risering did their jobs. Even when the designer of the casting does his job satisfactorily and the parts to be made are acceptable as potential castings, many decisions remain before a good, economical casting can be made. Pattern quality is usually a compromise between *economy and idealism*. One criterion of judgment in this respect is the number of castings to be made from the pattern. A few castings can often be made from a soft wood pattern before excessive wearing occurs; likewise, a costly hard wood or metal pattern cannot be amortized economically *for a limited number of uses*. An accurate, statistical appraisal can be made, but the pattern maker quite often decides this issue. A fair, general approximation is that the more expensive pattern usually proves cheapest in the long run. A carefully made, hard wood pattern will yield better castings than its soft wood counterpart because (1) moisture will not be absorbed as readily, (2) the impression in the sand will be sharper and more true to pattern, (3) less wear and erosion will accrue from repeated use, and (4) usually much more care and planning will be given a pattern made of expensive hard wood. Metal patterns are as much better than hard wood patterns as hard wood patterns are better than those made of soft wood, and for about the same reasons.

Sometimes blueprints are so complicated it is desirable to make small models of the casting to be made. The model may actually be used for making trial castings before the final pattern is laid out. This often leads to important changes in the full-scale pattern.

The skilled pattern maker must combine an ability to read a blueprint, visualize the finished product, and understand each step in the molding and casting process; he must bridge the gap between designer and founder, and must also have the skill to transmit what he envisions into a wood or metal replica of the final part. This requires no little skill, experience, and ability, and requires a working knowledge of mechanical drawing, and plane and solid geometry.

10.2 Materials for pattern construction

Patterns can be made from soft wood (northern white pine, sugar pine, Idaho pine), hard wood (mahogany from Central America, Mexico, and Peru), metal (brass, bronze, gray iron, aluminum, steel), plaster, plastics, and low-melting-point alloys (such as lead-bismuth alloys). For use in precision casting, expendable patterns are made of wax, tin, frozen mercury, and plastics.

Sometimes patterns are made from a combination of the various materials listed above. Metal inserts can be placed at locations on a wooden pattern which are subject to a high degree of wear. Plaster patterns can be impregnated with plastic resin to improve their strength and wear resistance. Combinations of materials are used for many other special purposes; some of these will be described later in this chapter.

10.3 Pattern taper and allowances for shrinkage, warping, and machining

As soon as the method of molding is determined, taper of not less than about $\frac{1}{16}$ to $\frac{1}{8}$ in. per ft is added to all surfaces essentially normal to the parting line; this is done to facilitate removal of the pattern from the tightly packed material of the mold without tearing or sticking. The average amount of taper varies with size and shape of pattern, and with molding material and method of molding. Figure 10.1 is an exaggerated example of pattern taper. It is not to be confused with so-called *design taper* (padding) used to facilitate feeding; this has been discussed in Chapter 6.

As a casting cools from solidification to room temperature, its

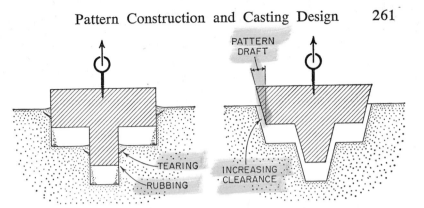

FIG. 10.1.   Example of pattern taper.   Taper is exaggerated for illustrative purposes.

over-all dimensions tend to become smaller.   Accordingly, it is necessary to compensate for this by making the pattern somewhat oversize.   The pattern maker uses a *shrink rule* (slightly oversize ruler), in constructing his pattern to accommodate this characteristic.   The amount of contraction varies slightly for different metals, and so different shrink rules are used.   Castings may not respond exactly to predetermined values of shrinkage; variations in mold resistance, use of hard cores, and design features such as casting size or shape, and the location of ribs, brackets, bosses, etc., and even the size and location of gates and risers, may prevent or alter normal shrinkage.   Although it is usually safest to rely on the shrink rule in the absence of a good reason to do otherwise, experience is always a valuable asset.   Table 10.1 lists typical shrinkage allowances for various types of castings.

Some correction for warping of the casting in the mold can be made in the pattern.   For example, a casting with long, rangy arms, like a propeller strut for a ship, usually warps with the arms tending to draw together.   By making the pattern outsize in the opposite direction, this dimensional warpage can be partially or totally corrected.   Experience is the only reliable guide for this type of compensation, although general rules are available for different shapes.

Castings often oxidize in the mold and during heat treatment, and one or more of their surfaces may be machined or ground to accurate size; also, it is sometimes necessary to remove surface

TABLE 10.1

Pattern Shrinkage Allowances *

Before specifying consult the pattern maker and foundryman.

Casting Alloys	Pattern Dimension, in.	Type of Construction	Section Thickness, in.	Contraction, in. per ft
Gray cast iron	Up to 24	Open		$\frac{1}{8}$
	From 25 to 48	Open		$\frac{1}{10}$
	Over 48	Open		$\frac{1}{12}$
	Up to 24	Cored		$\frac{1}{8}$
	From 25 to 36	Cored		$\frac{1}{10}$
	Over 36	Cored		$\frac{1}{12}$
Cast steel	Up to 24	Open		$\frac{1}{4}$
	From 25 to 72	Open		$\frac{3}{16}$
	Over 72	Open		$\frac{5}{32}$
	Up to 18	Cored		$\frac{1}{4}$
	From 19 to 48	Cored		$\frac{3}{16}$
	From 49 to 66	Cored		$\frac{5}{32}$
	Over 66	Cored		$\frac{1}{8}$
Malleable cast iron			$\frac{1}{16}$	$\frac{11}{64}$
			$\frac{1}{8}$	$\frac{5}{32}$
			$\frac{3}{16}$	$\frac{19}{128}$
			$\frac{1}{4}$	$\frac{9}{64}$
			$\frac{3}{8}$	$\frac{1}{8}$
			$\frac{1}{2}$	$\frac{7}{64}$
			$\frac{5}{8}$	$\frac{3}{32}$
			$\frac{3}{4}$	$\frac{5}{64}$
			$\frac{7}{8}$	$\frac{3}{64}$
			1	$\frac{1}{32}$
Aluminum	Up to 48	Open		$\frac{5}{32}$
	49 to 72	Open		$\frac{9}{64}$
	Over 72	Open		$\frac{1}{8}$
	Up to 24	Cored		$\frac{5}{32}$
	From 25 to 48	Cored		$\frac{9}{64}$ to $\frac{1}{8}$
	Over 48	Cored		$\frac{1}{8}$ to $\frac{1}{16}$
Magnesium	Up to 48	Open		$\frac{11}{16}$
	Over 48	Open		$\frac{5}{32}$
	Up to 24	Cored		$\frac{5}{32}$
	Over 24	Cored		$\frac{5}{32}$ to $\frac{1}{8}$
Brass				$\frac{3}{16}$
Bronze				$\frac{1}{8}$ to $\frac{1}{4}$

* From *Cast Metal Handbook*.[5]

TABLE 10.2

MACHINE FINISH ALLOWANCES AND BLUEPRINT MARKINGS *

Machine Finish Allowances—Various Metals

Pattern Size, in.	Allowances, in.		
	Bore	Surface	Cope Side
Cast iron			
Up to 6	$\frac{1}{8}$	$\frac{3}{32}$	$\frac{3}{16}$
6 to 12	$\frac{1}{8}$	$\frac{1}{8}$	$\frac{1}{4}$
12 to 20	$\frac{3}{16}$	$\frac{5}{32}$	$\frac{1}{4}$
20 to 36	$\frac{1}{4}$	$\frac{3}{16}$	$\frac{1}{4}$
36 to 60	$\frac{5}{16}$	$\frac{3}{16}$	$\frac{5}{16}$
Cast steel			
Up to 6	$\frac{1}{8}$	$\frac{1}{8}$	$\frac{1}{4}$
6 to 12	$\frac{1}{4}$	$\frac{3}{16}$	$\frac{1}{4}$
12 to 20	$\frac{1}{4}$	$\frac{1}{4}$	$\frac{5}{16}$
20 to 36	$\frac{9}{32}$	$\frac{1}{4}$	$\frac{3}{8}$
36 to 60	$\frac{5}{16}$	$\frac{1}{4}$	$\frac{1}{2}$
Non-ferrous			
Up to 3	$\frac{1}{16}$	$\frac{1}{16}$	$\frac{1}{16}$
3 to 8	$\frac{3}{32}$	$\frac{1}{16}$	$\frac{3}{32}$
6 to 12	$\frac{3}{32}$	$\frac{1}{16}$	$\frac{1}{8}$
12 to 20	$\frac{1}{8}$	$\frac{3}{32}$	$\frac{1}{8}$
20 to 36	$\frac{1}{8}$	$\frac{1}{8}$	$\frac{5}{32}$
36 to 60	$\frac{5}{32}$	$\frac{1}{8}$	$\frac{3}{16}$
Admiralty metal			
Up to 24	$\frac{1}{4}$	$\frac{1}{4}$	$\frac{3}{8}$

* From *Cast Metals Handbook*.[5]

roughness and/or imperfections. For these reasons, patterns are made slightly larger than the finished casting. Pattern-making manuals give dependably approximate values for machine finish allowances for different metals. Table 10.2 gives typical values.

10.4 Loose patterns, matchplates, cope and drag plates

Loose patterns are solid replicas of the castings to be made; they incorporate necessary core prints and allowances for metal contraction. Gates and risers are usually cut by hand, but these can also be constructed as loose pieces and molded with the casting. Molding a simple, split loose pattern is illustrated in Fig. 2.1.

A *matchplate* is used for production work and is only prepared when enough parts are to be made to justify the extra cost; its

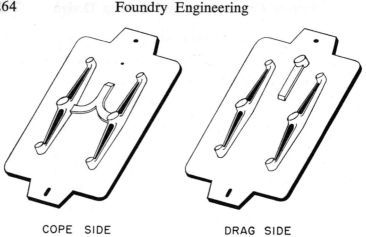

COPE SIDE   DRAG SIDE

Fig. 10.2. Matchplate pattern for casting two lever arms. Runner is
mounted on drag side of plate and gates on cope side; risers are not shown.

purpose is to obviate repeated handling of pattern parts and to
speed up production. In a matchplate the cope part of a pattern
is mounted on one side of a wood or metal plate, and the drag part
is mounted in accurate register on the opposite side. The parts of
the pattern forming the runner and gates are mounted on the
plate, and, whenever possible, riser patterns are also included.
Figure 10.2 is a sketch of a typical matchplate of the simplest type;
it has a horizontal parting line, or *flat match,* and is therefore
mounted on a flat plate.

When pattern shape is such that a flat match will not permit
drawing of the pattern, an *irregular match* must be used as shown
in Fig. 10.3. In this case the center of the plate is contoured to
accommodate the irregular parting line.

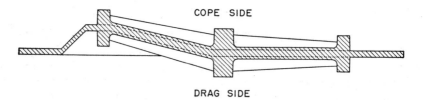

COPE SIDE

DRAG SIDE

Fig. 10.3. Matchplate pattern with irregular match. Section is through
pattern illustrating how plate is dished upward in the area of the pattern,
to form the irregular match.

Matchplates are most widely used for light work (when flask, mold, and pattern weigh less than about 100 lb). Molding with a rollover machine, as illustrated in Fig. 2.7, permits production of somewhat larger castings. Slip flasks, snap flasks, or tight flasks are used for matchplate molding. Slip and snap flasks are removed from the mold after ramming and used repeatedly; tight flasks remain with the mold until shakeout.

For heavier work than is readily handled in matchplate molding, or for increased production rates, patterns are mounted on *two* plates. The cope half of the pattern is mounted on one (*cope plate*) and the drag half on another (*drag plate*). This eliminates the necessity of turning over the cope and drag assembly several times during molding; it also permits two molders to work on opposite sides of a moving "line" of molds, one making drags and the other copes. Extra care must be taken to provide perfect register between cope and drag in this type of operation.

10.5   Pattern beds and follow boards

Solid loose patterns with no simple parting line can be molded with the aid of a *pattern bed* or a *follow board*. Pattern beds (sand matches) are beds of sand or other moldable material contoured to the natural parting line of a casting. Figure 10.4 illustrates the use of the technique in molding a ship's propeller. A flask is first rammed with sand, inverted, and the pattern embedded in the sand. Care is taken to cut and smooth the sand so the desired parting line is formed; this is the "pattern bed" (Fig. 10.4a). Figure 10.4b illustrates how the pattern bed would appear if the pattern were removed at this stage. In practice the pattern is not removed but the drag half of the flask is rammed over the bed; the mold is then inverted, the pattern bed discarded, and the cope half of the flask rammed. Remaining operations consist of removing the pattern, cutting gates and risers, and reclosing the mold as in molding with a simple split loose pattern (Fig. 2.1).

Follow boards are flat or recessed boards which perform much the same functions as pattern beds; they support the pattern in the first molding operations and form the natural parting line of the casting. Follow boards and pattern beds can be used interchangeably. If the pattern bed of Fig. 10.4b had been formed of wood, plaster, or metal, it would be termed a *follow board*. Molding sequences in using a follow board can be varied somewhat but are generally similar to those described for the pattern bed:

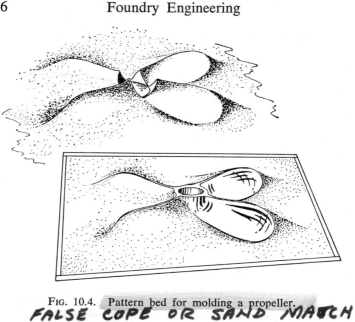

FIG. 10.4.   Pattern bed for molding a propeller.

FALSE COPE OR SAND MATCH

(1) The drag half of the flask is rammed around the pattern and follow board; (2) the mold is inverted and the follow board removed; (3) the cope is rammed. Subsequent operations are as in more standard molding. Figures 10.5 and 10.9 illustrate follow boards; molding with these is discussed in later paragraphs.

10.6   Drawbacks, cores, and loose pieces

Simple, solid patterns must be designed to "draw" readily out of the sand mold toward a parting line. To form internal cavities or projections which will not otherwise "draw," *drawbacks, cores,* and *loose pieces* are used. A sheave (pulley wheel) is an example of the type of part that cannot be readily withdrawn from a sand mold regardless of how it is parted; the rim of the wheel is "back-drafted" and will not draw (Fig. 10.5a). Sometimes sheaves are made laboriously by hand using *drawbacks* (sections of sand), which are pulled away after the cope is removed to allow the pattern to be drawn; the drawbacks are then returned to position. The cup, spoon, and saucer of Chapter 1 required such a technique, and Fig. 1.2 illustrates it in detail.

A simpler way to make the sheave would involve a core; also a core print would be added to the pattern to form a cavity for

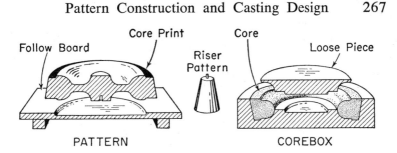

FIG. 10.5.   Patterns, and corebox for making a sheave casting by the baked-sand core method.

placing the core in the mold.   This is the more usual production method.   Figure 10.5 illustrates the pattern equipment required; if the rounded edges on the rim are desired, and if a solid one-piece pattern is employed, the follow board is necessary.   A cross section of the mold made with this pattern equipment is illustrated in Fig. 10.6.

Sometimes *green sand cores* are used to obtain internal con-figurations in castings such as the sheave.   Figure 10.7 illustrates one method of doing this, utilizing a split, but loose piece pattern. Note the extra rollover operation required if this method is used.

Loose pieces are used to form bosses or other configurations which will not "draw" with the pattern but which can be easily removed if drawn separately at an angle into the mold cavity.   An

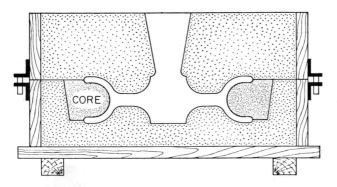

FIG. 10.6.   Cross section of mold for a sheave casting; baked-sand core method.

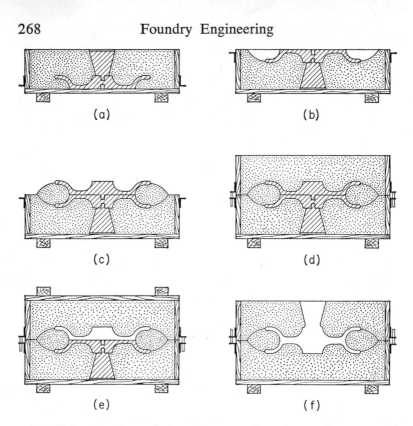

FIG. 10.7. Steps in producing a sheave casting using a split pattern and green sand core.

example is shown in Fig. 10.8. The boss could also be formed by a core; a core print would be added to the pattern to permit placing the core after molding (as in the case of the sheave), or the core might be rammed in place during molding. A less expensive alternative would be to change the design of the cast part so that the boss extended to the parting line and would draw without need of a loose piece or core.

10.7 Special molding techniques

Metal inserts are often positioned in molds or cores and formed integrally with the casting. If the surface of the metal insert protruding from the core into the mold cavity (i.e., the part around which the molten metal flows) is selected of proper metal, cleaned, and is dry at the time of pouring, and if the metal is poured at proper temperature, satisfactory fusion is obtained.

One of the little used molding techniques employs inflatable rubber forms molded to proper shape. These forms, or "patterns" of a sort, are inflated, positioned, and the molding medium is formed around them. The air can then be drawn out, and the pattern collapses and can be drawn from the mold. Complex runner systems, and unusual cavities, can be so made in the mold, and no problem is encountered in "drawing" the pattern.

10.8   Wood and metal patterns

The quality of lumber for pattern making is specified by the appropriate associations representing the soft wood and hard wood industries (Western Pine Association, Northern Pine Manufacturer's Association, Natural Hardwood Lumber Association). All wood for patterns should be thoroughly seasoned or kiln-dried and stored to prevent reabsorption of water. For ideal patterns, shop-seasoned lumber is preferred by most craftsmen, but this process requires from 6 months to 2 years; thus, most pattern

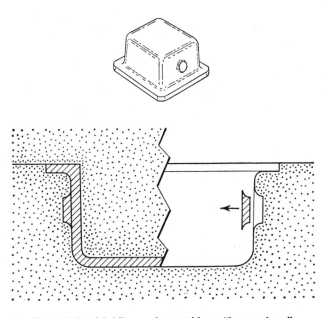

FIG. 10.8. Molding a boss with a "loose piece."

lumber used today is kiln-dried. It is extremely important to keep
moisture in pattern lumber at a low, constant value to prevent
warpage, swelling, and costly pattern repairs for dimensional correction during (or even before) use.

Loose patterns are usually of wood construction, although other
materials are becoming progressively more important. *Matchplate*
patterns are sometimes constructed entirely of wood, or they may
be wood patterns mounted on a metal plate; a more usual construction method is to make them entirely of metal as described
below. *Master patterns* are usually of wood construction; these
are original patterns which are used to cast one or more metal
patterns. Metal patterns so made can be loose patterns, mounted
on a matchplate, or can be cast integrally with a matchplate.

A pattern is cast integrally with a plate by first preparing a mold
of sand or plaster (from a master pattern) as though castings were
to be made. The cope and drag are separated by a suitable distance (¼ to ¾ in.) and a "dam" is provided between the outer
edges of the flask; metal is then poured through a sprue to the
cavity. The casting so made comprises a complete assembly of
patterns, gates, and risers cast integrally to a rigid plate in perfect
register.

10.9   Plaster patterns

Plaster of paris (gypsum) can be used to advantage in many
pattern applications; for example, it can be used directly for construction of the entire pattern, as matchplate material, or for
making models. It can be cut readily with saw or knife, formed
by scraping (screeding) with a sheet metal template, or turned
on a lathe, or can be molded directly to shape by pouring into a
mold; it can be easily patched, even after complete drying. Plasters
can be obtained that give patterns of relatively high hardness which
can withstand considerable usage. Bats of hemp, sisal, or twine
can be mixed with the plaster to impart toughness, if needed, for
rangy patterns. Also, plaster patterns can be *impregnated* with a
hard, tough plastic. Impregnation is accomplished by immersing
the pattern in a bath of plastic, applying a vacuum, and then re-
applying atmospheric pressure while the pattern is immersed.

Figure 10.9 illustrates *screeding* a plaster pattern with a template. In the pattern shown two templates were required, one to
form the follow board, and the other to form the top surface of

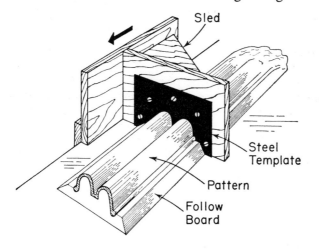

FIG. 10.9. Screeding a plaster pattern.

the casting.   The screeding operation is performed while the plaster is still in a plastic stage, before it begins to set.

10.10   Plastic patterns

Plastics have proved serviceable for pattern construction. They can be injected into a die to form a pattern or the pattern can be machined from a block of plastic of suitable size; the plastic can be cast into a plaster mold or it can be built up of successive layers of plastic and a fibrous material such as glass fiber. Until recently most plastic patterns were made by casting resins in plaster molds; today the method of laminated construction, building up successive layers of resin and glass fiber against a plaster mold is rapidly gaining favor. Phenol and epoxy resins are the two plastics most suitable for pattern work. Loose patterns and matchplate patterns can be made entirely of plastic, or plastic patterns can be attached to wood or metal matchplates.

Expendable patterns made from injected plastic are used for investment casting. Plastics are more costly and harder on dies than the wax more usually used for the purpose. Assembling several plastic patterns into a so-called "tree" is also more tedious than wax patterns which only require a hot spatula for the purpose. Advantages of plastic over wax for patterns for investment casting are better stability at room temperature and attendant better accuracy.   In spite of this, wax is more widely used.

10.11   Miscellaneous pattern construction techniques

Low-melting-point alloys, primarily lead-bismuth alloys, can be obtained with two characteristics which give them special value in pattern making: (1) they do not shrink on solidification or subsequent cooling, and (2) their melting temperature is so low that they can be cast in wood molds or against wood patterns.   The melting temperature of commercial alloys varies from 160 to 340°F (71–171°C).

Lead-bismuth alloys are used to produce cast patterns which are exact replicas of patterns in service.   If materials such as aluminum are used for this purpose, the cast patterns are slightly smaller than the original, owing to metal contraction.   To produce duplicate patterns in all but the low-melting, non-shrinking alloys requires that a *master pattern* be constructed.   The master pattern is slightly larger than the final cast pattern (it is constructed with a "double shrink rule" to compensate for shrinkage in both the cast pattern and final casting).

Another application of low-melting alloys is the "keying" of wood, metal, or plastic patterns in matchplates by flowing a low-melting alloy into a matching groove in pattern and plate.   Patterns for such assemblies are often interchangeable by melting the metal in hot water or a low-temperature furnace and reassembling. Fillets can be added to wood patterns by gouging a "key" in the pattern and casting the alloy against the pattern (in a suitable mold) to form the exterior of the fillet; runners and gates can be formed on matchplates in the same manner.   Other uses of low-melting alloys are described in the literature.[1, 2]

Sometimes patterns are made by *electroforming*.   A plaster (negative) pattern is first formed and a thin layer of copper electrolytically deposited.   The copper forms the pattern surface; it is usually backed up with plaster or a low-melting-point alloy to achieve the necessary strength.

10.12   Pattern colors

Standard colors have been recommended for the finishing of wood patterns.   The color scheme adopted by the American Foundrymen's Society is outlined below:

(*a*) Surfaces (of the casting) to be left *unfinished* are painted *Black*.

(*b*) Surfaces to be *machined* are painted *Red*.

(*c*) *Seats of and for loose pieces* are *Red Stripes on a Yellow Background*.

(*d*) *Core prints* and *seats for loose core prints* are painted *Yellow*.

(*e*) *Stop-offs* * are indicated by *Diagonal Black Stripes on a Yellow Base*.

The pattern code is used strictly by some foundries and not at all by others. Also, with the present widespread use of (unpainted) metal and plastic patterns, the color code is less widely used.

10.13   Summary of pattern construction

Patterns influence the quality of castings made from them to such degree that the most expensive pattern is often cheapest in the long run. For long production runs, metal patterns yield the best molds; pattern, risers, and runner system should be mounted on a *matchplate* or *cope and drag plates*.

Materials other than metals such as aluminum and steel for making patterns include hard wood, plastics, metallized wood, plaster, soft wood and low-melting-point metals. Patterns made from these materials are *loose patterns* (one piece or split) or patterns mounted on *matchplates* or *cope and drag plates*. Molding aids, primarily to permit drawing intricate patterns from sand molds, include *pattern beds, follow boards, drawbacks, cores,* and *loose pieces*.

Patterns should never be made without due regard to the feasibility and economics of the casting method. Also, the method of pattern construction will be determined by (1) number of castings to be produced, (2) storage or "shelf" life required of the pattern, and (3) degree of quality required in the final casting.

10.14   Introduction to casting design

Many castings are literally doomed to failure on the drafting board. Some designs should never be accepted by the foundry-

---

* Stop-offs are portions of a pattern that form a mold cavity which is filled with sand before pouring. Stop-offs may, for example, be reinforcing members to prevent breakage of a frail pattern.

man if he is interested in making a quality casting. Once the design is fixed it is difficult to obtain alterations or concessions; the time for collaboration between designer and foundryman is *before,* not *after,* design is decided upon. Care at this stage saves time, money, and castings.

There are definite limitations to the foundryman's ability to make *sound* castings if the original design is incorrect; unless the design engineer has considered the potential casting from the viewpoint of foundry requirements, the end results may be uneconomical or unsatisfactory. It is for this reason that a designer or consumer of castings needs experience in pattern making and foundry operations, or should seek guidance from pattern makers and foundrymen before final designs are developed. With complicated castings it is often desirable to make a model to determine accurately whether standard molding methods can be used and whether it is possible to obtain sound metal in all critical areas.

10.15   Predesign considerations

Before committing himself to action on the drawing board, a design engineer should ask himself: (1) Is the design such that the pattern can be adapted for economical molding by standard methods? (2) Can gates, risers, and chills be positioned properly to insure soundness? (3) Are the section size and configuration such as to cause undue stresses in the mold and consequent tearing or cracking? (4) Can directional solidification be established and controlled, or (5) should the casting be broken down into component parts and the separate castings welded together? (6) Is it possible to delete certain members such as feet, bosses, and unwieldy protuberances, and weld them to the casting later to make foundry operations easier, cheaper, and more dependable?

It is possible to carry these considerations so far that the inherent advantages of the casting process are not fully realized; the proper engineering balance can be determined only by knowing and applying minimum requirements for good foundry practice.

10.16   Design for minimum casting stresses

Because of high pouring temperatures, ferrous castings are particularly susceptible to external cracking or tearing in the mold; total contraction is high, and at critical temperatures hot strength and ductility are low. These defects may be caused by influences external to the casting or by inherent characteristics.

The only external influence of importance is the effect of the mold, which may restrain normal contraction and cause cracks, tears, or warpage. Stresses can be reduced by molding expedients, as discussed in the preceding chapter, or by *design considerations*. If the casting is large, and of complex shape, it may prove desirable to cast two or more sections and weld the parts together. Small *tie bars* are sometimes used to prevent tearing in critical areas, and straight members, like spokes of wheels, may be curved to reduce stresses slightly.

Stresses arising from inherent conditions are those due strictly to design and those peculiar to the composition of the metal. Thin sections of castings cool and contract much faster than heavier members. If such members are attached rigidly, high stresses develop; as the smaller members cool and shrink, they are restrained by larger, slower-cooling sections. The small members may either tear to relieve the stresses or deform plastically without tearing; if they do not tear, warpage occurs when the larger members contract away from the extended smaller parts.

Tearing may even develop in straight bars not restrained at the ends. As the section cools and contracts, enough resistance may develop from friction between mold and metal to rupture the casting. If the bar is uniform in cross section, no particular region of weakness will develop and the bar will not tear because solidification and contraction occur uniformly across the section. However, if another member is joined to the straight bar, a hot spot is formed and the solidification pattern is conducive to hot tearing.

Metal composition influences tearing tendencies in at least three ways: (1) by the inherent strength and ductility at critical temperatures; (2) by the existence and extent of solid transformations; and (3) by the presence of impurities, such as sulfur, at grain boundaries. No extensive quantitative data exist for the guidance of design engineers. In general, high-quality, plain, low-carbon steels resist tearing better than alloy steels, and gray irons are less susceptible than alloy irons. In iron and non-ferrous castings tearing is usually more a matter of design than of composition.

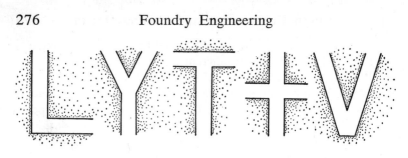

Fig. 10.10.   The basic forms of casting junctions.

10.17   Design for directional solidification

It has been established that temperature gradients in solidifying castings must be favorably controlled if sound castings are to be made. The design engineer is very largely responsible for the relative ease or difficulty with which this can be done by the foundryman; sometimes design is so bad that adequate directional progressive solidification cannot be attained. In that event, redesign is essential. For easy understanding, the various design considerations will be itemized and discussed in order.

1. *Heavy sections cannot be fed through light sections.* Various members or component parts of castings are joined in L, Y, T, X, or V fashion (Fig. 10.10). The T section is selected to represent the series. By inscribing circles, as in Fig. 10.11, it is easy to determine that the region depicted by circle *d* is a larger mass of metal than those of circles *a*, *b*, or *c*. This means that, under normal conditions of cooling, metal at the center of the *d* region

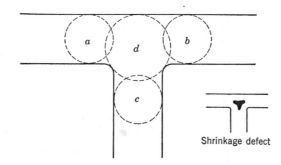

Fig. 10.11.   Method for the determination of hot-spot location by inscribing circles.

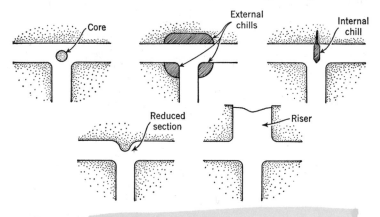

FIG. 10.12.   Methods for eliminating hot spot in a T section.

is a *hot spot,* the last to solidify.   Usually it is neither convenient
nor economical to place a riser at this point, and it is necessary to
rely upon feeding through one or more of the *arms* from a riser
placed some distance away.   Since the arms are smaller, they will
naturally freeze before the junction area, thus *cutting off* feeding,
and shrinkage will develop.   Figure 10.12 shows various corrective
expedients.

The designer should attempt to limit junctions to as few as
possible and to select the least difficult; the L joint, for example,
is less troublesome and easier to treat than Y or V sections; the
L section is also easier to correct by design alone than any of
the others.   Sometimes it is impossible to limit the number of
junctions except by casting in segments and joining these together
later by welding or bolting; this expedient is often very worth
while.

2. *If possible, sections should taper toward risers.* The importance of this principle has been clearly established.   Often
some compromise must be made with the ideal condition to meet
weight and cost requirements.   Within reason, casting soundness
obtained by natural solidification is proportional to the degree of
tapering allowed.   Figure 10.13 illustrates the design of a valve
casting: one-half shows the *usual* situation in which no attempt
has been made to enhance feeding by natural methods; the other
half is correctly designed.

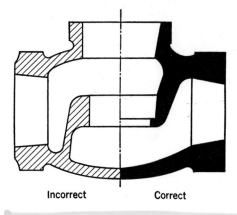

Incorrect   Correct

FIG. 10.13.   Designing a casting for controlled directional solidification.

Cylindrical castings are difficult to make sound by static methods; they are best made by centrifugal casting. To make cylindrical castings (of steel or non-ferrous metals) free from center-line shrinkage by static methods, tapering or some more elaborate expedient is necessary. The cylinder may be tapered and risered in any of the three ways shown in Fig. 10.14.

3. *Horizontal flat surfaces are not desirable.* Castings with flat surfaces are particularly difficult to make because (1) it is hard to prevent center-line shrinkage, as temperature gradients are not favorable, and (2) slag, dross, and other impurities lighter than metal tend to collect on upper flat surfaces. Whenever pos-

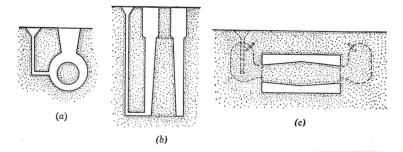

(a)

(b)

(c)

FIG. 10.14.   Alternative methods of padding a tubular casting to promote directional solidification.

sible, it is best to have curved surfaces or to make castings so that flat surfaces are either vertical or inclined to the horizontal.

4. *Isolated hot spots must be avoided.* Small cores or pockets of sand surrounded by metal reduce the rate of cooling as compared to more open regions, and such isolated areas are potentially troublesome. Without special means for chilling or feeding, these regions are likely to shrink or tear.

10.18 Design for metal flow

Minimum section thicknesses for sand castings are a function of metal composition. In general, sand castings of gray iron and aluminum, magnesium, and copper-base alloys should not be designed in less than $\frac{1}{8}$-in.-thick sections in very small castings, nor to less than $\frac{3}{16}$ to $\frac{1}{4}$ in. in larger castings; steel castings, if small, should never be less than $\frac{1}{4}$ in. in thickness, and, if large, less than $\frac{1}{2}$ in. Irons containing 0.5 to 1 per cent phosphorus (so-called stove-plate irons) are sometimes as thin as $\frac{1}{16}$ in.; household radiators are an example. These considerations are conservative and apply only to average sand foundry practice; new casting methods greatly extend the field for thinner castings, but they are special developments.

10.19 Cast-weld design

Cast-weld structures present many advantages. The tendency of a casting to warp or tear is proportional to its length and complexity. By separating a casting into parts and welding the components together, or welding forgings and castings together, the versatility of all the fabricating processes is enhanced; this method of fabrication is more economical than all-cast or all-welded construction, and sounder castings can be made. The use of cast-weld structures is increasing rapidly.

10.20 Safety factors

Safety factors employed in casting design can often be decreased. Engineers usually employ excessively high factors of safety, often as high as 3 to 1, whenever castings are for critical structural applications. The mechanical properties of sound castings, properly made from well-melted metal, are comparable to those of wrought materials. Unfavorable comparisons are often made between

forgings and castings on the basis of parts that are not as amenable
to casting as to forging. With inspection methods currently available, and with improved foundry practice, there is no need to
employ a faulty casting in a strategic application. A capable
foundry engineer knows whether a particular casting is likely to
be equivalent to its wrought counterpart.

10.21 Design for low pattern cost and minimum manipulation

After blueprints have been studied, and design is well fixed in
mind, there still remains the annoying problem of pattern construction. Most design engineers leave this to the pattern maker, who
may or may not consult the foundryman before going ahead. The
problems of pattern making should be understood at least, before
design is fixed, so as to determine if anything can be altered to
make pattern construction more economical. It is too much to
expect a designer to be a gifted pattern maker, and comprehensive
treatment of the subject is beyond the scope of this book. Practical experience is again the best guide as it is not possible to
formulate rigorous rules; but it seems desirable to emphasize the
importance of design in pattern costs. The number of operations
needed to make molds and cores and to assemble them for a casting
is determined by design; also much can be done on the drawing
board to minimize the number and complexity of cores needed.
Core making and core setting are expensive operations.

For most design engineers their problem ends when the final
tracing goes to the blueprint room; it would be an excellent idea
if they could follow their creations through all the steps to the final
casting, and make design changes wherever such changes would
favorably affect the cost or quality of the end product.

10.22 Model making as an aid in design

Trouble and money would be saved for producer and consumer
alike if complicated castings could be made as an accurately scaled
wood or plastic model; even the best-trained engineer or foundryman misses important details of design when all he sees is a blueprint. Many costly design changes and innumerable casting repairs could be obviated by making and studying a model of the
job to be done. Areas of potential shrinkage and/or tearing can

be predicted, gates and risers can be judiciously placed without trial and error, and money-saving changes in pattern construction can be spotted before working patterns are made; if necessary a model-sized casting can be made, and cut up for inspection, to determine potential casting problems.   By making sectional models, each part of the casting can be studied by an entire committee and important decisions made.   Transparent plastics are excellent for model work; modeling clay can be used to simulate padding and other important features.   Enough models have already been used to prove the desirability of this approach.

10.23   Summary to casting design

Good casting design is rapidly becoming appreciated as a very important requirement for making satisfactory castings.   For this reason, *a designer or consumer* of castings needs experience in pattern making and foundry practice, or should seek guidance from trained specialists before final designs are fixed.   He should ask himself: (1) Does the design permit economical manufacture by standard methods?   (2) Can risers and other appurtenances such as gates and chills be positioned to insure soundness?   (3) Can potential cooling stresses be minimized by simpler design?   (4) Is directional solidification enhanced naturally?   (5) Can the casting be made better as a cast-weldment, and can any feet, bosses, flanges, and other protuberances undesirable from the standpoint of potential soundness be deleted?   (6) Has maximum advantage been taken of the flexibility of the casting process?

Among the specific rules for satisfactory design are: (1) design for *minimum casting stresses* by limiting drastic changes of thickness; (2) design for directional solidification by tapering sections as liberally as possible toward risers; (3) do not interpose thin sections between the risers and heavy sections; (4) avoid extensive horizontal, flat surfaces; (5) avoid isolated *hot spots* which are difficult to feed; (6) determine casting properties and use commensurate safety factors; (7) simplify complex castings by dividing into natural components and welding them together; (8) use minimum section thickness commensurate with the flowing qualities of the metal designated; (9) be sure that specifications are realistic and sensible, and, finally, (10) *do not compromise on the rules;* remember that, the more the engineer can help the foundryman at the design stage, the better will the foundryman be able to make the castings required.

10.24 Problems

1. You are required to produce a small, aluminum casting (with cores) to close dimensional tolerances. What type of pattern equipment might you use for (*a*) 3 castings, (*b*) 100 castings, (*c*) 10,000 castings?

2. List three advantages of a matchplate pattern over a loose piece pattern.

3. What are the advantages of plastic patterns over those made from metal? What are the disadvantages?

4. What is the difference between pattern taper and design taper? What is a flat match?

5. If a pattern does not have a "flat match," what techniques can you use to mold it and be able to draw the pattern?

6. List five steps you might take to eliminate hot tearing in a steel casting.

7. Is design for directional solidification as important in aluminum alloy and gray iron castings as it is in steel castings? Why?

8. What steps might you take to assure yourself (in the early design stages) that a particular part will be designed well for casting?

9. List four rules that must be considered in design of castings.

10. Why are castings designed on a minimum-casting-stress basis?

11. What must be controlled to obtain directional solidification? Mention three ways in which this can be attained.

12. List a number of different alloys, and classify them with respect to minimum castable section thickness.

13. Show by sketch how internal and external hot tears appear in radiographic examination. Explain the intrinsic differences between them.

14. Obtain small patterns of metal and wood (identical if possible) from a local foundry. Place these in a flask and ram identically. Remove the patterns and observe the mold cavity carefully. Differences should be apparent in mold cavity smoothness and possibly in moisture retention unless the wooden pattern is made of hard wood carefully prepared with several coats of shellac. Scrape the shellac from a small area, or use an unshellacked piece of smooth wood; then repeat and see how worn patterns affect mold perfection.

15. Ask a local pattern maker for a blueprint of a complicated casting. Study this and you will appreciate why a model is always helpful and sometimes essential. Some of our larger industrial concerns

like Westinghouse and General Electric would be pleased to send such prints on request.

10.25   Reference reading

1. *Pattern Maker's Manual,* A.F.S., Des Plaines, Ill., 1953.

2. *Cerro Alloys in Foundry and Pattern Shop,* Technical Bulletin, Cerro De Pasco Corp., New York.

3. *Design of Gray Iron Castings,* Gray Iron Founders' Society, Cleveland, 1956.

4. *Product Information, Data, Applications and Use,* U. S. Gypsum Co., Chicago.

5. *Cast Metals Handbook,* fourth edition, A.F.S., Des Plaines, 1957.

CHAPTER **11**

Storage, Handling, Preparation
and Reclamation of Sands

11.1 General considerations

It is becoming increasingly evident that methods used for storing sands have a direct bearing on the quality of castings and on the ultimate cost of the finished product. Foreign matter and excess water, for example, collecting upon sands stored in open piles or pits preclude adequate control. Sand handling methods are as important to a successful foundry as is materials handling in any industry. Some foundries handle ten carloads and more of sand each day, and so a saving in time and costs is a significant item. Dependable production of good castings naturally hinges on proper preparation of molding and core sands, and reclamation is strictly a matter of economy which depends heavily on the size of foundry, amount of new sand used per day in relation to total sand used, and the type, quality, and size of castings produced. Methods available for preparing, handling, storing, and reclaiming sands are many, and improvements are being made steadily.

11.2 Preparation of molding and core sand mixes

The oldest, and perhaps the simplest, method of preparing sand for molding is by hand shovel. This involves pouring a quantity of water over a heap of used or new sand and systematically working the moisture throughout the mass. Perhaps from this crude process evolved the so-called "mulling" action inherent in mechanical mixers (mullers) used today in the most modern shops. To do a thorough job of mixing by shovel the worker would spread a shovelful of sand from the heap in a thin, neat pile on the floor —he would then press the back of the shovel down upon the sand

284

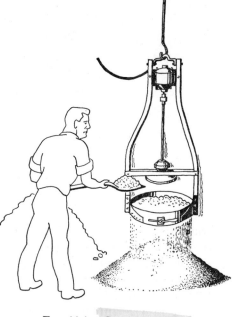

FIG. 11.1. Gyratory riddle.

and draw it across the mass. After the heap had been "worked" through two or three times in this manner the clay of the sand was tempered and ready for use. One of the first improvements on this procedure was use of a screen (forerunner of our modern aerators) through which the sand was passed before use (Fig. 11.1). Hand mulling was successful because all the sands used in early days were of the so-called "natural" type; the "synthetic" sand used so widely today cannot be readily or properly tempered in this manner as bentonite clays require considerably more energy to develop plasticity than the illite and kaolinite clays with which natural sands are bonded.

A definite improvement over hand mixing was use of a rotary sand kicker which straddled long heaps of sand, previously moistened with water, and mixed the mass with vigorously rotating blades as the machine moved along the row (Fig. 11.2).

Modern sand mixers, or "mullers," operate on one of two general principles in which tempering and mixing are accomplished:

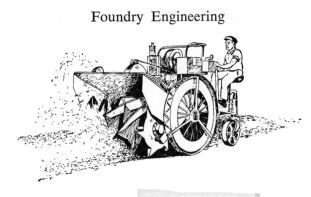

Fig. 11.2. Rotary sand kicker.

(*a*) with thick rimmed, heavy, metal wheels rotating relatively slowly over the moistened mixture of sand and clay, or (*b*) with lighter, thinner, rubber-covered wheels spinning rapidly over and through the mass. In the former type the wheels are always rotating with their axes horizontal and the full weight of the wheels resting on the sand; the wheels may move with the pan held stationary (Fig. 11.3), or the wheels may be mounted rigidly with the pan rotating (Fig. 11.4). In either case the wheels are not power-driven but are rotated by friction against the sand. Plows keep the sand stirred and in position under the wheels.

Fig. 11.3. Sand muller, stationary pan.

FIG. 11.4.   Sand muller, rotating pan.

In the high-speed unit the wheels rotate with their axes vertical and the sand is stirred vigorously against the vertical sides of the mixer shell (Fig. 11.5).   A small portable unit is available which is a compromise between these two types.   In this unit all working parts are mounted into a cover which fits snugly over a specially shaped wheelbarrow.   After mixing, the unit is lifted from the wheelbarrow by crane, and the mixed sand can be wheeled wherever it is needed in the shop and another batch prepared in a second wheelbarrow (Fig. 11.6).

FIG. 11.5.   High-speed muller.

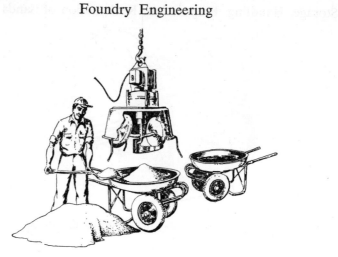

FIG. 11.6. (Portable muller.)
MULL - BARROW

So-called screenerators are often used as mixers. After use, sand is moistened and shoveled upon a vibrating screen, through which the sand falls upon a rapidly spinning wheel or belt (Fig. 11.7). The sand is mixed and thrown from the belt or wheel into a pile as desired, usually into a corner of the shop or against a three-sided backboard.

Whatever type mixer is used, care must be taken to remove large, hard lumps of sand and metal objects. Equipment for this purpose is sometimes simple, such as vibrating screens, or elaborate, when magnetic separators, crushers, and vibrating screens are incorporated into a unit.

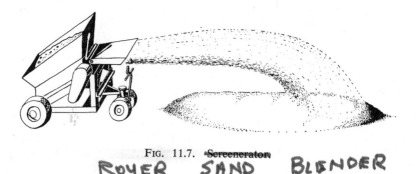

FIG. 11.7. Screenerator
ROYER SAND BLENDER

Core sands can be mixed in any of the above units or in pug-mills or bakery-type mixers. To avoid contamination, core sands and molding sand are usually mixed in separate units.

The fundamental purposes of mixing are considerably different between core and molding sands. In core sands it is only necessary to coat the sand grains as uniformly as possible with oil and water, or with resin; in molding sand the energy of mulling is required to plasticize the clay (and hasten the absorption of water by the clay).

A typical mulling cycle comprises addition of a weighed amount of sand to the mixer together with predetermined amounts of clay and other dry ingredients. These are mixed dry for about 2 minutes (in the slow-speed mixer), after which water is added and mulling continued for about 5 minutes. Times are correspondingly shorter for the high-speed muller.

11.3   Handling molding sands

Handling new sand as it arrives at the foundry from the producer will be discussed under storage, as the method of handling is influenced by storage facilities.

Sand is taken from the muller in batches; each batch is caught in a portable container of suitable size or is dumped into a bucket elevator or upon an endless belt. As soon as enough batches are collected, the portable container is moved by hand or motor truck or by overhead conveyor to the molding station. Sand containers should always be covered tightly and made of metal to prevent loss of moisture in the tempered sand. In the more mechanized shops endless belts deliver the sand to each molding station where it is scraped into overhead hoppers or stationary bins. In smaller, less mechanized shops, the sand may be taken from the muller in wheelbarrows; if so, the tempered sand should be covered with a moist bag to prevent surface drying.

11.4   Sand storage

Sand storage has always been an annoying problem for foundries. Since an average amount of sand required to make a ton of steel castings, for example, is from 1 to 2 tons it is obvious that sand comprises the largest single item of raw material used. This is an exaggerated example, perhaps, since less sand is used per ton of non-ferrous castings than for steel, but sand is always an item of major concern. It is not possible to operate with low sand inventory because freight shipments might be held up and, in winter, sand is often not mined or shipped in extremely inclement weather. Also, unless sand is stored in some form of shelter, it will often be wet or frozen and will always contain variable amounts of moisture, quite often higher than can be tolerated. Sometimes several different grades or types of sand are required by shops, which again complicates the problem.

Sand normally arrives in railroad boxcars or gondolas, or by truck; it may be in bags but, unless ordered otherwise, will nearly always be in bulk. It is unloaded by crane or endless belt, or by

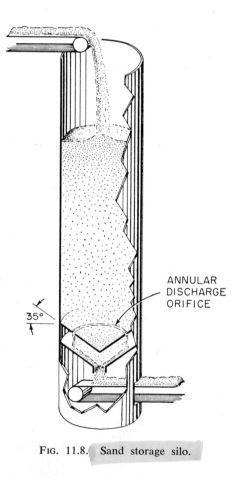

ANNULAR
DISCHARGE
ORIFICE

35°

Fig. 11.8. Sand storage silo.

wheelbarrow, tote truck, or bucket truck. It is placed either in open piles or bins, or inside or outside in silos; the silo is rapidly becoming the popular means for storing sand as so much can be stored in a small floor area. One problem in connection with silo storage, in particular, is the tendency of sand to segregate in fine and coarse fractions as it is removed. This problem can be solved by designing the bottom exit of the silo properly. Figure 11.8 shows a design which has proved satisfactory.

Sand is taken from the silo by wheelbarrow, tote truck, bucket elevator, or endless belt to the weigh hopper of the muller and dumped automatically or manually.

11.5   Reclamation of sands

Reclamation of sand is the process of restoring used foundry sand to as nearly its original condition as possible. The methods for doing this are generally classified as (1) dry reclamation, (2) wet reclamation, (3) thermal reclamation, and (4) combined wet (or dry) reclamation plus thermal reclamation. Figures 11.10 to 11.12 depict these processes schematically, and their general operational characteristics will be described.

Sand reclamation is desirable only when the cost of the operation does not greatly exceed the cost of new sand, and it surely must not exceed the cost of sand plus freight charges; also, it is indicated only when sand is used in relatively large quantities and is principally applicable to synthetic-type sands. In natural sands complete reclamation, in which all clay and residue is washed from the grain, would be prohibitively expensive in terms of wasted clay (which may be as high as 12 to 18 per cent compared to 3 to 4 per cent bentonite in synthetic sand). Sand reclamation is particularly applicable to steel foundries where much new sand must be used for facing each new mold. To show the enormity of the problem, one shop (not the largest) produced 20,000 net tons of large steel castings in one year for which they purchased 50,000 tons of new sand. Since all the sand entering a shop must eventually leave it to make room for new sand purchased, disposal is a problem. For a shop of this sort there are very impelling reasons for reclaiming sand:

1. Properly reclaimed sand is equivalent in performance to new sand in all respects.

2. Finding places to dump huge quantities of used sand is a problem, and loading and hauling is expensive.

3. Storage of new sand in quantities necessary to accommodate such a requirement is expensive and troublesome, and huge inventories are required, particularly in the winter months.

4. Even though the base cost of sand is relatively low, freight costs are extremely high.

5. By reclaiming all sand, there is less tendency to "skimp" on the use of new sand facing to the detriment of casting quality.

6. The troublesome task of unloading cars of sand is obviated, since storage of reclaimed sand is automatic.

7. Sand is available at constant moisture content and of consistent quality (particularly grain size and grain-size distribution).

8. It is necessary to understand clearly the physical need, as well as the economic need, for sand reclamation in order to appreciate how the various methods of reclaiming accomplish their mission, and why some items of equipment, such as scrubbers and aerators are needed. Again, the case for sands used in casting steel will be taken as the example because conditions are more severe than for the lower-melting-point metals. The conditions for sands used for cast iron would be similar but not identical.

To impart such properties as green and dry strength, collapsibility, permeability, and flowability and to modify mold-metal interface reactions or to adjust grain-size distribution, such materials are added as fireclay, bentonite, lignins, cereal binders, linseed and proprietary oils, starches, resins, iron oxide, wood flour, and others. These materials form a coating around each sand grain which varies in thickness with the amount added and with the number of times the sand is used without reclamation. Figure 11.9 illustrates the layers of clay which are built upon sand grains in this manner.

The sand grains at and near the mold-metal interface undergo physical and chemical changes due to the intense heat. This results in drying, burning, and baking of the materials surrounding the sand grains into a hard, tightly adhering coating which may or may not be soluble in water; some of the sand grains themselves also react with iron oxide at the surface of the casting to form fayalite ($2FeO \cdot SiO_2$). Since active clay is rendered inactive if its water of crystallization is removed, a considerable amount of this material also exists, which adds to the so-called "fines" and

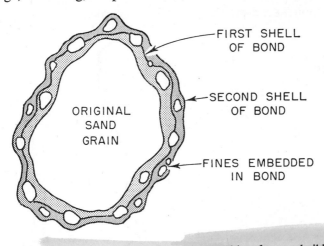

FIRST SHELL
OF BOND

SECOND SHELL
OF BOND

ORIGINAL
SAND
GRAIN

FINES EMBEDDED
IN BOND

FIG. 11.9.   "Shell" on a used sand grain, resulting from a buildup of successive layers of bond, with fines embedded.

contributes nothing when the mass is remulled; in fact, additional active clay must be added to bond this material.   In core sands some of the already hard shell of baked linseed oil bakes even harder.   There is some fracturing of sand grains due to heat shock but this is probably unimportant.   Also, there is an inevitable accumulation of iron and steel and iron oxide particles of various shapes and sizes.   It is obvious that this conglomerate mixture will not be suitable for reuse at the mold-metal interface since it would be of doubtful composition and much less refractory than new silica sand.   The fine materials seriously reduce permeability and the impurities reduce refractoriness.   It took many years for steel foundrymen to realize that new sand (or reclaimed sand) facings should always be used at the casting interface, and some are still not convinced of this.   At any rate it is clear by now why it is desirable to remove the non-silica fraction from used molding sands.

Dry reclamation removes fines such as silica flour, spent and free clay, finely fractured sand grains, and iron oxide particles; usually iron and steel particles are removed by magnetic separators, and lumps are crushed and screened.   Simple dry reclamation accomplishes its mission by screens and air separation; if a pneumatic scrubber is included, the tenacious coatings on the sand grains are

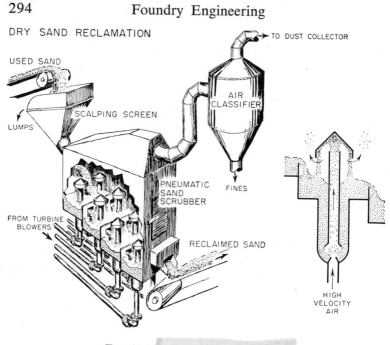

FIG. 11.10. Dry sand reclamation.

removed, at least in part. It is not possible to remove the clay coatings from sand grains unless individual grains are rubbed over and against one another repeatedly at high velocity. Dry reclamation alone does not restore the sand to its initial quality since some clay and also organic and carbonaceous materials may cling to individual sand grains.

Figure 11.10 illustrates one type of dry reclamation system. Iron and steel particles are first removed by magnetic separation (not shown) and the used sand is fed to a scalping screen. Lumps that are left on the screen may be discarded or crushed and re-screened. The sand that passes through the screen enters the pneumatic sand scrubber unit where the grains are repeatedly picked up and rubbed against each other to scrub off the coatings. The sand grains then travel to an outlet while fines and dust are separated from the air.

Wet reclamation is similar to dry reclamation in that its purpose is to remove fines and foreign materials, and to clean the individual sand grains of tenacious coatings. Complete removal of clay coatings can be obtained in well-designed wet reclamation units; wet units have thus far proved more effective than dry units in removal of these coatings. Wet reclamation does not, however, restore sand to its original quality; some carbonaceous and organic residues remain on the individual sand grains.

Figure 11.11 illustrates a wet reclamation system. Used sand is first passed through a magnetic separator and screens, as in dry reclamation. It is then mixed with water to form a *slurry,* passed to a *primary classifier* to remove fines and to a wet storage tank. The sand-water slurry next enters one or more *sand scrubbers* which rub the sand grains against one another to remove the adherent clay. The sand scrubber is simply a tank with a large propeller immersed in it; the propeller rotates at high velocity in the sand-water slurry, creating the "rubbing" action. Next, the slurry moves to a secondary classifier to remove the fines created in the cleaning operation, and also to obtain the desired sand grain-size distribution. Finally the water is removed from the sand by a filter and drying unit.

Thermal reclamation comprises heating used sand to 1200–1500°F (650–800°C) in furnaces which are usually of the multiple hearth type. This treatment removes carbonaceous and organic materials effectively (by burning) but has little effect on the clay coatings of sand grains. The heat from the furnace may, in fact, make these coatings bond still more firmly to the clay. Thermal reclamation units can fully reclaim oil-bonded sands which have no clay bond. Simple thermal units have been used with success in several non-ferrous shops which perform predominantly all-core molding, and they are also effective for reclaiming shell mold sand. Even in these specialized cases, thermal reclamation should be combined with screens and classifiers to assure proper grain-size control of the used sand. By itself, thermal reclamation is not adequate for most shops because it does not remove clay.

By combining *thermal reclamation* with *wet reclamation,* the benefits of each are realized, and it is possible to reclaim foundry sands to their original quality. Fines, adherent clay, organic and carbonaceous material are all removed. The sand is restored to its original grain size, shape, and distribution, and the castings made from such reclaimed sand are fully as good as those made in new sand. Except for a slight iron stain and some slight pitting

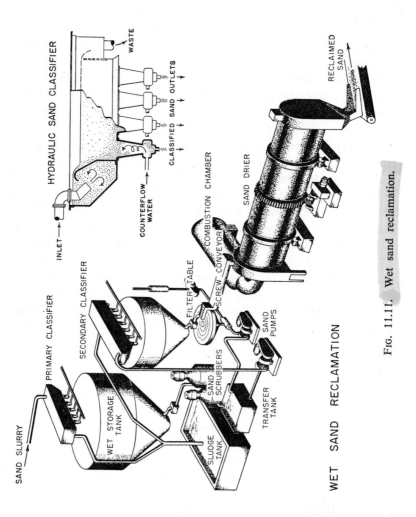

Fig. 11.11. Wet sand reclamation.

and etching of the individual grains, the restored sand, after thermal treatment, even looks like new sand.

Figure 11.12 illustrates a combined wet-thermal system for complete sand restoration. Components of the wet system shown are similar to those described earlier. The used sand, after passing through the wet portion of the system is classified, dewatered, and then enters a rotary (multiple) hearth furnace for the thermal treatment at 1200 to 1500°F (650 to 850°C). The sand is finally cooled, reclassified and stored.

Standard sand reclamation systems of all the above types are available, or it is possible to custom-design a reclamation unit and buy separate parts of the system from different suppliers.

11.6 Summary

Sand *storage* and sand *handling* methods are as important to a successful foundry as is materials handling in any industry. Dependable production of good castings hinges on proper *preparation* of molding and core sands. Sand *reclamation* is strictly a matter of economy; the cost of reclaiming old sand must be balanced against the cost of using new sand for the same purpose.

Molding sands, especially synthetic molding sands, require a considerable amount of "working" in the mixing operation to develop plasticity in the clay bond. To perform this work, mechanical mixers (*mullers*) are now used in most modern shops. Other sand-mixing methods in use include simple *hand shoveling, riddling,* and use of a *"sand kicker,"* or *screenerator.* The various types of mechanical mixers are illustrated in Figs. 11.1 to 11.7.

Sand storage has always been an annoying problem for foundries. Variable moisture content, segregation of fines, and freezing in the winter are some of the difficulties encountered. To combat these problems, the *silo* is rapidly becoming a popular means of storing sand.

After each use, the quality of molding sand deteriorates unless effective *sand reclamation* is employed. The deterioration is much more rapid in the case of steel than in the lower-melting nonferrous metals; it is due to a variety of causes including (1) increase in "fines," (2) buildup of clay and organic coatings on sand grains, and (3) presence of foreign materials such as iron particles and fayalite ($2FeO \cdot SiO_2$). The purpose of sand reclamation is to remove fines, foreign materials, and tenacious coatings on the sand

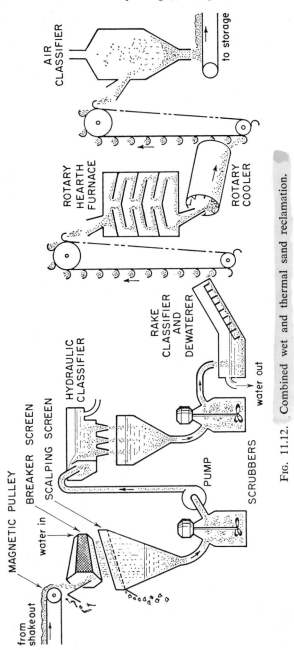

FIG. 11.12. Combined wet and thermal sand reclamation.

grains—in short, to restore the sand to as nearly its original condition as possible.

Reclamation methods which are discussed in this chapter are (1) *dry reclamation,* (2) *wet reclamation,* (3) *thermal reclamation,* and (4) combined *wet reclamation plus thermal reclamation.* The last method is the only one of the above that *completely* restores original sand quality.

11.7   Problems

1. List the advantages and disadvantages of each of the types of sand mullers.

2. Why is hand tempering satisfactory for natural sands but unsatisfactory for synthetic sand?   Prepare samples of each type of sand with a shovel, and compare the green compressive strengths obtained with samples mixed in a muller.   If a muller is not available, use a rubber bag and knead the latter samples thoroughly; if a green compression tester is not available, make one, using simple levers and a bucket into which shot is poured until the specimen breaks.

3. How does the silo of Fig. 11.8 act to prevent sand segregation?

4. What factors would you consider in deciding whether or not to adopt a complete wet thermal sand reclamation unit in a steel foundry?

5. What type of reclamation system would you expect to see in a large aluminum foundry doing predominantly green sand molding?

6. Why is "scrubbing" an essential part of any sand reclamation system?

7. Why is it so important to use new or reclaimed sand as a facing for molds in a steel foundry?

8. What advantages might reclaimed sand have over new sand from the standpoint of foundry operations?

9. Obtain a bag of used, black sand from a local steel foundry. Simulate the various steps in a wet and dry reclaiming system as described in this chapter, using pails, Gooch funnels, coking ovens or Bunsen burners, etc., and study the sand after each step with a microscope; try some ideas of your own—you might invent a better method.

10. Call a local foundry and ask the names of three foundry sand suppliers; then write to these companies and obtain literature describing their mining, separation, blending, and shipping processes and their various grades of sand.   After studying this material select five completely different grades of sand from at least one company and ask for samples of these.   After studying these grades under the microscope and checking their green properties (if the necessary sand testing equipment is available), write the company again and arrange for a visit to the plant; you will be more than welcome and

you will see a far more interesting operation than you would ever imagine, particularly if you choose the most modern and progressive sand supplier.

11. Make a collection of all different sands (molding sand, glass sand, shot blast sand, just plain sand, etc.), place them in small bottles, label clearly and place in a cupboard; then write a short thesis on the genesis of sands and describe their chemical and physical structures.

12. Do the same for clays of all types.

13. Fill a porcelain dish from the chemistry laboratory half full of clean silica sand; place a 1-in.-long piece of plain carbon-steel drill rod (about ½ in. or less diameter) on top of the sand, and then cover by filling the dish with more sand. Do the same, using a similar specimen of stainless steel. Place both dishes in a heating furnace at 1500 to 2000°F for 1 hour; remove specimens and observe the condition of the pin surface and the sand lying against the pin. Repeat the experiments at increased temperatures if necessary until considerable interface reaction has occurred in the case of the plain carbon-steel pin. This is a good way to simulate mold-metal interface behavior when molten metal is poured into sand molds.

14. Do the same, using inert atmospheres and/or vacuum; air-conditioning molds to control interface reactions are a definite possibility for the future. This is essentially what is done when sulfur-bearing compounds are mixed with sand for casting magnesium—otherwise magnesium would catch fire and burn.

15. If facilities are available, make a more critical comparison (using X-rays, petrographic microscope, etc.) between interface reactions obtained in a laboratory dish using solid specimens with those obtained when sand specimens are immersed in molten metal.

11.8 Reference reading

1. Dietert, H. W., "Processing Molding Sand," *Transactions A.F.S.,* **62,** 1954.

2. *Foundry Sand Handbook,* A.F.S., 1953.

3. Dietert, H. W., *Foundry Core Practice,* A.F.S., 1950.

4. *Symposium on Foundry Sand Reclamation,* A.F.S., Des Plaines, Ill., 1953.

CHAPTER 12

Melting

PART I. MELTING

12.1 Introduction to fuels and furnaces

A complete treatment of fuels and furnaces would be a valuable contribution to the literature but would be prohibitively extensive for this text book. Far too little attention is paid by foundrymen to this important phase of foundry operations. In this chapter, only enough space can be devoted to the subject to describe the various melting units as they are constructed and used for melting metal for casting purposes. Figures 12.1 to 12.8 present the essential features of construction of each unit.

All types of metal-melting furnaces can be used for foundry operations; however, foundry requirements are sometimes unique and one or another type of furnace may be best for a particular operation. For example, open-hearth furnaces are desirable in large foundries requiring 10 to 200 tons of molten metal at one time. In a smaller jobbing shop, where 1 to 15 tons of metal may be needed at frequent intervals, an electric-arc furnace may prove best. In still another foundry requiring only 250 lb to 1 ton of alloys of various special compositions, an induction furnace will give best all-around performance. The choice of furnace may be dictated by (1) considerations of initial cost, (2) relative average cost of maintenance and repair, (3) base cost of operation, (4) availability and relative cost of various fuels in the particular locality, (5) cleanliness and noise level in operation, (6) melting efficiency, in particular the speed of melting, (7) degree of control (metal purification or refining) required, (8) composition and melting temperature of the metal (for example, a cupola is normally used for melting cast iron but seldom, if ever, for melting

TABLE 12.1

Fuel	Form of Fuel	Metal Melted	Furnace
Coal	Powdered bituminous,	Cast iron	Air furnace
	anthracite lump, or	Cast iron	Cupola
	briquettes	Cast iron	Cupola
Coke	Sized lump	Cast iron, non-ferrous	Cupola, crucible furnace
Oil		Non-ferrous, iron, steel	Crucible furnace, open-hearth
Gas		Non-ferrous, iron, steel	Crucible furnace, open-hearth
Electricity	Direct arc	Steel, cast iron	One-, two-, and three-phase electric-arc furnace
	Indirect arc	Non-ferrous, cast iron	Rocking arc furnace
	Resistance	Non-ferrous	Rocking arc furnace
	Induction	All metals	Tilting and lift-induction furnace

aluminum or magnesium), and (9) personal choice or sales influence.

Fuels or power sources for melting metals, and the types of furnace with which each is normally used, are listed in Table 12.1.

12.2   Crucible furnaces

The earliest forms of clay vessels for holding molten metals were similar to the crucible currently used (Fig. 12.1). This vessel was placed in a pit dug into the ground; wood was packed around the crucible and ignited. Draft was provided by bellows or chimney to burn the wood and melt the metal in the crucible. Figure 12.1b is a standard pit-type furnace; several crucibles can be placed in a single pit. Today coke, oil, or gas can be used as fuel. All non-ferrous metals are melted readily in coke-fired units, and cast iron can be melted if enough time is allowed. Steel can be melted in crucibles in pit furnaces using oil- or gas-fired units; in fact, very excellent steels have been made in this manner. (The steel for making the famous Damascus swords was melted in pit crucible furnaces.) Today crucible melting is used chiefly for non-ferrous metals.

The pit furnace is rapidly being replaced by crucible furnaces operating at floor level (Fig. 12.1*c*). No expensive chimney is needed for draft, and crucible handling is easier and safer. The coke-fired, forced-draft furnace (Fig. 12.1*d*) is a quiet, economical unit for melting non-ferrous metal; to date only relatively few such furnaces are used in spite of their apparent utility.

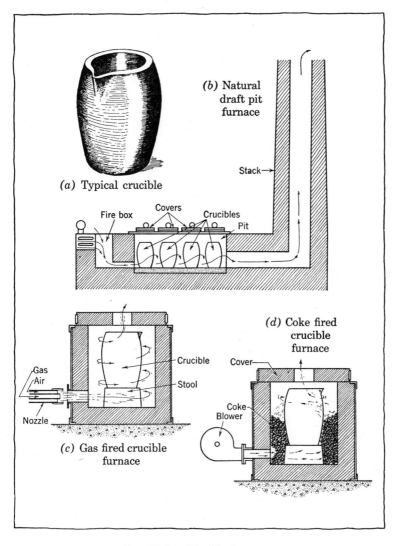

FIG. 12.1. Crucible furnaces.

12.3   Arc furnaces

The *rocking arc* furnace (Fig. 12.2a) is used predominantly for melting copper-base alloys. An electric arc is drawn between two electrodes above the metal charge (so-called *indirect arc*), and

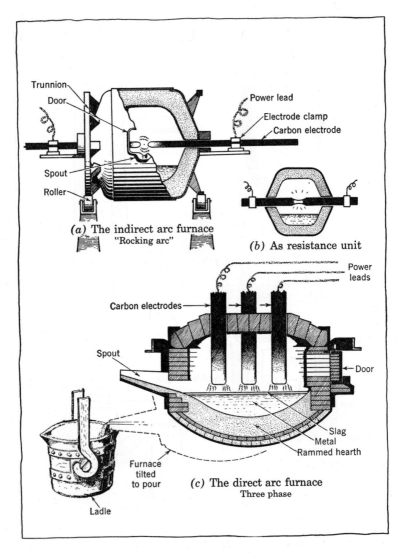

(a) The indirect arc furnace
"Rocking arc"

(b) As resistance unit

(c) The direct arc furnace
Three phase

Fig. 12.2.   Electric-arc furnaces.

heat radiated to the furnace interior melts the metal. A rocking motion of the furnace keeps the bath stirred and insures uniform temperature and composition of the metal. Cast iron and steel can be melted in this unit, but for such use melting efficiency is low and refractory costs high.

The rocking arc furnace can be operated by resistance heating of a single electrode. Reduced diameter of the central portion of the solid electrode (*b* of Fig. 12.2) induces greatest resistance to passage of current in this region and furnishes heat for melting. This unit is completely noiseless, but operational cost is high and few such furnaces are used.

The three-phase electric-arc furnace (Fig. 12.2*c*) is the most popular unit for melting steel in the foundry, although total tonnage of steel produced for castings compares about equally with that melted in open-hearth furnaces. Three-phase furnaces vary from 1 to 100 tons' capacity; the most popular units are 3 to 8 tons' capacity—only about two 100-ton and perhaps a dozen 25-ton furnaces are in use in America. In operation, scrap steel is charged into the furnace through one of two doors, or *top-charged* if the roof section is removable as it is in newer units. An arc is drawn directly between each electrode and the charge beneath (so-called *direct arc* type) to furnish heat for melting. The arc is regulated by automatic controls which raise or lower electrodes to maintain desired arc voltage. Slags are maintained on the molten bath (1) to reduce oxidation, (2) to refine the metal, and (3) to protect roof and sidewalls from excessive heat radiated from the molten metal and the arc. The melting rate is relatively rapid, and control over temperature and composition of metal is excellent. When melting is complete and the metal is refined and/or superheated to the desired pouring temperature, the electrodes are raised and the metal is emptied into ladles by tilting the furnace.

12.4   Open-hearth furnaces

Open-hearth furnaces (Fig. 12.3) are used by a few large foundries and by nearly all shops producing ingots. The melting rate is relatively slow, depending upon furnace size, and 10 to 200 tons may be melted at one time. Scrap metal, pig iron, and slag-forming materials such as lime or sand are charged into the furnace through side doors by large charging *rams*. Heating is done with oil or gas; fuel is fired through nozzles working alternately for about 20-minute periods from opposite ends of the furnace. Air for burning oil or gas is preheated by being drawn over brick checkerworks at each end of the unit; the checkers are heated alternately by passing hot waste gases from the furnace

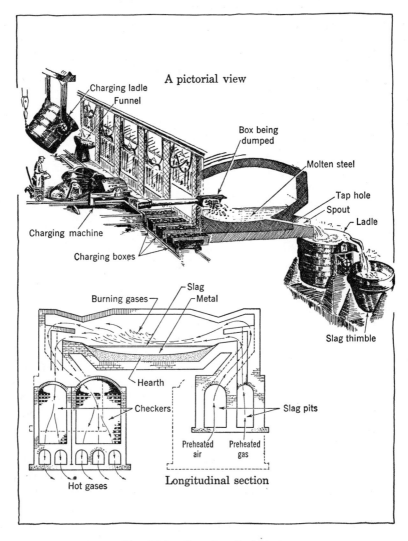

Fig. 12.3.   Open-hearth furnace.

MEMORANDUM

For _____ 19___

<u>Melting Point</u> - temp at which
metal begins to liquify.

Fluidity - ability of molten metal to
flow freely as measured in inches
in the spiral channel

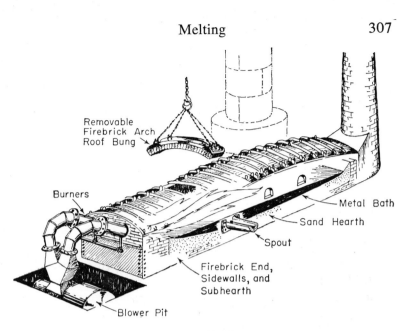

Fig. 12.4. Air furnace.

over them while the burner at that end of the furnace is idle. This is known as the *regenerative* system for preheating air for combustion of fuel. The metal bath in the open-hearth furnace is extremely shallow to provide the greatest practical surface area for a given volume; this improves heat transfer and increases reaction rates between metal and the slag used for refining the metal. Small open-hearth furnaces can be tilted for pouring, but large units are *tapped* by opening a *taphole* at the bottom level of the bath; the taphole is located on the back at the midspan of the furnace. This description applies to open-hearth furnaces for melting steel, as this is their chief tonnage use. They are equally useful for melting large quantities of aluminum, copper, and alloys of these metals.

The *air furnace* (Fig. 12.4) is essentially of open-hearth type. The unit is fired at one end only, using powdered coal for fuel; air for combustion may be preheated in regenerators as described above. Fuel (bituminous lump coal) is pulverized by high-speed rotary crushers and blown into the furnace as fine dust; oil may also be used as fuel. The products of combustion are drawn across the metal charge and out a draft stack. Air furnaces are charged by removing sections of the arch roof, called *bungs,* and dropping the scrap, pig iron, or other ingredients through the hole; the bungs are then replaced and sealed with clay.   Air furnaces are used in the foundry chiefly for melting iron; some are for gray iron, but most units are designed for making white iron for malleableizing.

12.5   Induction furnaces

Induction melting is relatively new for foundry use.   Many types and sizes of induction melting units are available; the most versatile type is the *coreless induction furnace* (Fig. 12.5).   Some coreless furnaces operate at 960 cycles and others at frequencies ranging as high as 100,000 cycles or more.   A very recent development is the 60-cycle coreless furnace which operates directly from line current.   The principles of operation are essentially the same in all units.   In Fig. 12.5 the melting unit proper is a simple shell of transite board within which hollow, water-cooled copper coils are mounted.   The coils are covered with wet refractory dried into a hard mass.   The melting lining is then formed or placed inside the coils.   One method is to pack dry refractory around a metal cylinder which is melted with the first charge; an alternative method employs a prefabricated crucible located centrally inside the coil and packed into place by ramming dry refractory tightly between the crucible and precoated coils.   Melting is by resistance of the metal of the charge to passage of secondary current induced in the charge by electromagnetic induction; the coils are supplied with current from a motor-generator, spark-gap converter, or, in the case of the 60-cycle unit, directly from line current with a capacitor bank to balance power factor.

Any metal can be melted in an induction furnace.   The charge is placed inside the crucible and current is applied; more metal is charged as the melting proceeds.   One of the chief advantages of the induction furnace is its ability to melt relatively small quantities of an unlimited variety of metals and alloys cleanly, conveniently, and quickly; stirring of the bath, caused by induced lines of force (*c* of Fig. 12.5), insures excellent homogeneity of composition. *Tilting* and *lift-coil* units are available (*a* and *b* of Fig. 12.5).

Although the 60-cycle furnaces are still very new, it appears that they will prove the most economical of the coreless furnaces for continuous or semi-continuous melting in the foundry.   They are efficient, and melt at a rapid rate if a heel of molten metal content

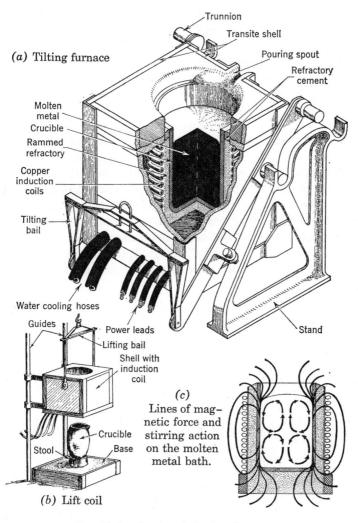

(a) Tilting furnace

Trunnion
Transite shell
Pouring spout
Refractory cement
Molten metal
Crucible
Rammed refractory
Copper induction coils
Tilting bail
Water cooling hoses
Power leads
Stand

Guides
Lifting bail
Shell with induction coil
Crucible
Stool
Base

(b) Lift coil

(c)
Lines of magnetic force and stirring action on the molten metal bath.

FIG. 12.5. Coreless induction furnaces.

is left in the furnace after tap. Stirring action is greater in the 60-cycle unit than in the higher-frequency units, and, when a liquid heel is present, stirring quickly melts down additional charges of even, very finely divided, scrap. A disadvantage of the unit is that, if it is completely emptied and "cold" metal melted, the initial metal charge must consist of relatively large ingots or billets. Higher-frequency furnaces easily melt small pieces of "cold" metal (scrap, master alloys, etc.); they are used in preference to the low-frequency furnaces for rapid melting of small heats, and when operations require frequent changes in the metal to be melted and poured.

The 60-cycle *cored induction furnace* (Fig. 12.6) is the most efficient of the induction melting units. Here, the induction coil is essentially immersed within the metal bath. Electromagnetic induction pumps liquid metal through channels around the coil; simultaneously secondary currents (which act as the heating source) are induced in the liquid metal around the core. The major disadvantage of the cored induction furnace is that it requires a liquid metal "starting" charge; it therefore is not suitable for intermittent operation. The cored furnace is used primarily for

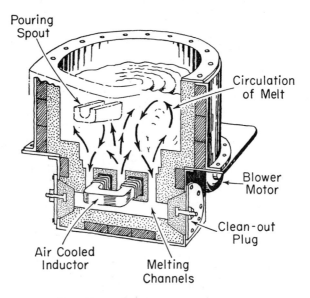

FIG. 12.6. Cored induction furnace.

non-ferrous metals in applications where relatively long periods of continuous operation are desirable.

12.6   Cupolas

The cupola is the standard melting unit of the iron foundry; cupola melting is generally considered the cheapest method for converting cold scrap metal or pig iron to usable molten iron. The features of the cupola (Fig. 12.7) are essentially those of a typical blast furnace, well known for use in converting iron ore to pig iron.   Fuel for the cupola is preferably a good-grade, low-sulfur coke; anthracite coal or carbon briquettes may also be used. The furnace is charged at frequent intervals with proportioned amounts of coke, limestone, and metal (pig iron, scrap steel, or scrap iron).   Melting can be continuous over long periods of time, or intermittent, as required.

In operation the cupola is essentially a shaft-type furnace lined with refractory bricks.   To prepare for melting, *bottom plates* are swung upward to form the bottom of the unit and are held in place by a bar or pipe.   A sand lining is rammed over the plates, and a *taphole* is formed at this level in line with the tapping *spout;* wood and excelsior are placed on the sand bottom and covered with several feet of coke.   The unit is now ready for *firing.* Excelsior is ignited, and the coke *bed* eventually becomes red hot.

Several *charges* (lime, coke, and metal in balanced proportions) are dumped through the charging door upon the coke bed, and at the proper time the *air* blowers are started.   Air is forced from the *wind box* through *tuyères* into the furnace.   This air, rising upward through the *stack,* furnishes oxygen for combustion of coke; air may or may not be dried and/or preheated.   Besides being fuel the coke supports the charge, or *burden,* until melting occurs.   The limestone melts and forms a flux which protects the metal against excessive oxidation; lime also fuses and agglomerates the coke ash.   As melting proceeds, metal collects at the bottom of the shaft; it may be *tapped off* at intervals, or the taphole may be left open with metal flowing constantly.   With open taphole the rate of melting and the size of taphole must be in balance.   In most cupolas slag is drained from a *slaghole* at the back of the furnace; in some installations *front slagging* spouts are used and slag flows out with the metal to be separated from it in a *fore hearth* or in the ladle.   When melting is complete, the *bottom bar*

is pulled sharply from under the plates and the *bottom is dropped.*
All remaining slag, unburned coke, or molten metal drops from
the furnace.   When the unit has cooled it is patched and made
ready for the next *heat.*

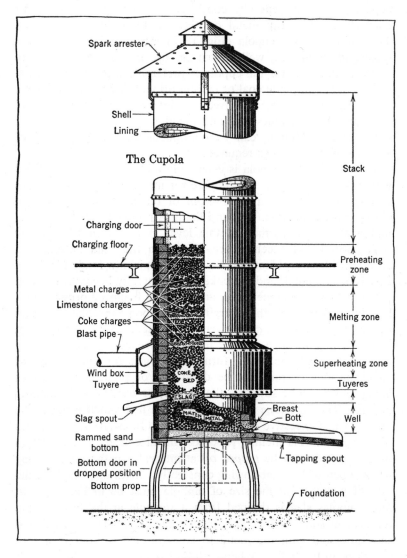

The Cupola

Fig. 12.7.   Cupola.

12.7 Converters

Cast iron, containing the usual amounts of carbon, manganese, and silicon, can be *converted* into steel by *blowing air through or over* molten iron; this was one of the first European production

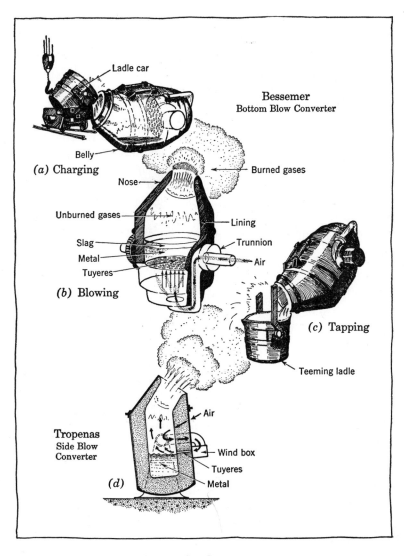

Fig. 12.8. Converters.

methods for making steel. In operation molten iron is transferred from the cupola by ladles to an oval vessel (Fig. 12.8) called a *converter.* The vessel is leveled, and the iron is poured into the open *nose;* as soon as the unit is tilted back into position air is blown through *tuyères* in the bottom or side (*a* and *b*, respectively, of Fig. 12.8). This air oxidizes silicon, manganese, and carbon. High-carbon cast iron is reduced to very low-carbon iron essentially free of silicon and manganese; the blow is continued only until carbon is reduced to about 0.10–0.20 per cent, to prevent undue oxidation of iron. The order of oxidation of elements and their rate of oxidation can be followed by the color and length of flame issuing from the converter nose; originally the *end point* of blow was determined by eye, but more recently photoelectric controls have been developed. No fuel is required for *converting;* the oxidizing reactions are *exothermic,* and more heat is added to metal than is removed by air passing through or over the charge. Ferroalloys of carbon, manganese, and silicon are added to the molten iron as it is tapped into a pouring ladle, to adjust the metal to required composition. Contrary to some opinions, converter steel can be of as high quality as steel made by other methods.

PART II. MELTING PRACTICES

12.8 Introduction to melting practices

The many factors which dictate choice of a furnace for a specific melting practice were outlined in the first section of this chapter; the characteristics and operation of the furnaces were subsequently described. It is now our purpose to discuss some general principles of the *melting practices* which are carried out in the various melting furnaces.

The foundry melting operation usually involves much more than simply adding enough heat to a solid metal to liquefy it. Care must be taken in charging and melt-down operations to avoid damage to the furnace refractories. The melting operation, particularly the slag condition (if a slag is used), and the furnace atmosphere, must be carefully controlled to assure that alloying elements are not lost, and that the metal does not dissolve unwanted impurities. Often the metal must be *alloyed* during melting by addition of other metals. When undesirable impurities are present in the initial charge, or are absorbed during melting, they must be removed by *refining*. One type of refining, the removal of gaseous impurities, is usually termed *degassing* (Chapter 5). At the end of the melting operation, metals are sometimes *inoculated* to improve their final solidification structure. Such inoculation treatments are to *grain-refine* the metal, to produce *ductile iron,* or to alter the metallurgical microstructure in other ways.

12.9   Metals and alloys used for melting

The form of melting stock (metals used for making alloys that go into castings) varies with the end product and, to a lesser degree, with the type of furnace used for melting. Good melting stock is often a strategic commodity; one of our most critical industries during World War II was so-called *junk collecting*. Often the quality of castings is influenced by the selection of melting stock for the charge.

Non-ferrous alloys are nearly always melted from *primary* or *secondary* ingot, and from foundry *returns* (gates, risers, and scrapped castings from previous operations). When the alloy composition must be adjusted, pure alloying *elements* are added if the melting point of the elements are sufficiently low. If the melting point of an alloying element is too high, a lower-melting-point *master alloy* (alloy of the element and base metal such as copper plus 15 per cent phosphorus, or aluminum plus 5 per cent titanium) is used to make the charge addition. Alloying elements are also sometimes added from a *salt* of the element (for example, a flux such as titanium chloride).

Cast iron is made from pig iron, and iron and steel scrap. Foundry returns are also used. Alloying elements are generally added as master alloys of iron plus the alloying element (*ferroalloys*). Often there is an economic advantage in the use of ferroalloys as compared with alloying with pure metals.

Primary ingot is prepared from *commercially pure* metals by refineries known as smelters; secondary ingot is a remelt of *scrap metals* sorted, melted, and alloyed by refiners to approximately the composition required by the foundryman for his castings. Secondary ingot is cheaper and less strategic than primary ingot— and sometimes less desirable. Scrap may consist of railroad rails, worn, obsolete, or broken machinery and miscellaneous castings, or loose or baled borings, turnings, *flashings* from razor blades and other wrought objects, etc.

12.10 Refining

Refining in the melting operation is the removal of unwanted impurities from the metal bath.   The general types of impurities that may be present are (1) dissolved gases, (2) dissolved non-gaseous elements, and (3) suspended oxides and inclusions.

In non-ferrous melting, pig is generally purchased of the proper chemical composition for casting.   No attempt is made in the foundry to remove dissolved impurities (except dissolved gases). Dissolved gases are removed by one of the degassing techniques described in Chapter 5.   A refining operation to remove entrapped oxides and inclusions is often necessary in non-ferrous metals— especially in aluminum and magnesium alloys.   These alloys easily form solid oxides, called *dross,* which tend to remain suspended in the melt.   Sufficient time must be allowed for the particles to float to the surface in the furnace and ladle, or they must be *fluxed* from the metal by stirring a reactive salt flux into the bath.

Refining in the ferrous foundry, particularly the steel foundry, is more extensive than in the non-ferrous shop.   In the case of steel, two general types of melting practices are used, *acid* practice and *basic* practice.   Acid melting involves the use of acid refractories and acid slags in contact with the metal; such slags and refractories are high in SiO_2 content and low in CaO and MgO. Basic melting involves slags and refractories of relatively high CaO and MgO content.   In *acid* melting, dissolved gases and suspended non-metallics are removed, oxidizable impurities are partially or completely removed, and carbon composition is adjusted.   Phosphorus and sulfur contents are not, however, affected, and, since these impurities are deleterious to the mechanical and casting properties of steel, acid melting requires an initial charge low in these elements. *Basic* melting possesses the metallurgical advantages of acid melting, with the addition that sulfur and phosphorus may be reduced to any desired level by properly controlling the many melting variables involved, particularly the concentration of lime and iron oxide in the slag.   Basic melting, therefore, can employ a lower and cheaper grade of scrap as charge material.

In the melting of ferrous metals, phosphorus removal is promoted by slags which are both *basic* and *oxidizing* (high in CaO, MgO, and FeO).   Sulfur removal is promoted by slags which are *basic* and *reducing* (high in CaO, MgO; low in FeO).   Thus, melting conditions for optimum sulfur removal are not the same as those for optimum phosphorus removal. For example, melting of steel in the basic open-hearth furnace takes place under oxidizing conditions; the basic open hearth reduces phosphorus contents of steel to very low values but is less efficient in regard to sulfur. In the basic cupola, melting of cast iron takes place under reducing conditions; sulfur is readily removed but under ordinary melting practices little reduction of phosphorus takes place. In electric-furnace melting of steel, both phosphorus and sulfur may be reduced to very low values by a "double slag" melting practice. The metal is melted under a highly oxidizing basic slag to accomplish phosphorus removal; this slag is then removed and replaced with a basic reducing slag to lower the sulfur content.

Until recently, refining operations in cast iron were limited. Essentially all cast iron was melted in the acid cupola, a furnace which permits close control of carbon and silicon contents, but does not remove sulfur or phosphorus. The advent of ductile iron has required that a more "pure" grade of iron be produced in large tonnages (notably, low sulfur iron). Sulfur elimination in cast iron is now accomplished by (1) melting in a basic cupola, (2) melting in a basic electric furnace, or (3) melting in an acid cupola and adding desulfurizing compounds in the forehearth or ladle. Since phosphorus removal is not ordinarily accomplished in the cupola, it is necessary to avoid high-phosphorus charge materials for the cupola melting practice; both phosphorus and sulfur may, however, be removed by a double-slag electric-furnace operation. Desulfurization in the ladle may be accomplished in a variety of ways, two of these being to pour over a sodium oxide slag or to "inject" powdered desulfurizers such as calcium carbide. Injections is accomplished by blowing the powders (with an inert gas) into the metal through a tube immersed in the melt.

12.11 Inoculation

A wide variety of inoculants are presently used in the foundry. Most of these have been developed only in recent years, and it appears quite probable that many more will be discovered, and will find commercial application. An *inoculant* is an addition made to a melt, usually late in the melting operation, which alters the solidification structure of the cast metal. Some common inoculation treatments include *grain refinement* of aluminum and magnesium alloys, *graphitization* for gray cast iron, and the use of magnesium or cerium to produce *ductile iron*.

In *grain-refining* aluminum alloys, less than 0.2 per cent titanium or 0.02 per cent boron is sufficient to reduce the cast grain size of the alloy from as much as 0.10 in. in diameter to as little as 0.005 in. in diameter. Magnesium-aluminum alloys are grain-refined by small additions of carbon to the melt, or by superheating the metal to about 1650°F. In this respect, magnesium-aluminum alloys are somewhat anomalous; most other cast metals tend to exhibit a coarser grain structure when they are superheated high above their melting point. Magnesium alloys which do not contain aluminum are usually grain-refined with small additions of zirconium.

A slight different type of inoculation treatment is the graphitizing inoculation of gray cast iron. Here, the inoculant promotes graphite formation and is used to prevent "chilling" (formation of white iron in thin sections), and to avoid the undesirable "interdendritic graphite" structure which sometimes forms on rapid cooling. Common graphitizers employed are ferrosilicon, nickel-silicon, silicon-manganese-zirconium, and calcium-manganese-silicon. These inoculants must be added late in the melting operation or their effectiveness is much diminished.

A most important and very new inoculation treatment is that of adding small quantities of magnesium (or cerium) to cast iron to produce ductile iron. The microstructural alteration accompanying this treatment was described in Chapter 4. The addition of as little as 0.04 per cent residual magnesium alters the graphite flakes to almost perfect spheroids, with tremendous improvement in mechanical properties of the alloy. A metallurgically similar phenomenon to the magnesium treatment of ductile iron is the sodium inoculation of aluminum-silicon alloys. In this case, minute quantities of sodium (or phosphorus) alter a needle-like Al_3Si precipitate, to a finely divided eutectic structure. This treatment has been in use for many years; it substantially improves the ductility of aluminum-silicon alloys containing more than about 8 to 10 per cent silicon.

The theory of inoculants is as yet incompletely understood. It is known, however, that the action of these additives is due to more. than a simple alloying effect. Two melts with analytically identical final chemical compositions may produce very different micro-structures, depending on the *time, temperature,* and *type* of inoculating additions. Typical of this is the graphitization inoculation of cast iron; a ferrosilicon addition made late in the melting operation is a far more effective graphitizer than a similar addition made earlier (even though the final chemical compositions are the same). The effects of inoculants seem to be due to more subtle causes than chemistry; for example, grain refiners and graphitizers probably promote nucleation in the melt by introducing "foreign" nuclei into the liquid metal. Other inoculants, such as those that produce ductile iron may change the surface tension of solidifying particles and thereby alter nucleation and/or growth.

12.12   Metal handling

Handling molten metal comprises the various operations from the time the "melt" is "tapped" from the furnace until it has been "poured" (sometimes called "teemed") into molds. Sometimes molten metal is poured directly from the furnace into a mold or molds, without subsequent transfer by ladles. Such practice is the exception rather than the rule.

Sometimes metal is transferred by ladle from the melting unit into a large holding furnace, and thence by smaller ladles into molds. This provides greater flexibility and better utilization of furnaces and blends the metal from several units into a more uniform product.

The following discussion will be concerned only with the more standard methods of metal handling applicable in the average shop and will not present details of special methods used by one or a few large shops, although these methods are important.

Almost without exception molten metal is handled carelessly in one or another respect. Sometimes safety regulations are practiced scrupulously but cleanliness is completely ignored, and sometimes accent is on the latter but safety is ignored. It is particularly important to handle molten metal with care and respect. Besides the safety hazard, molten metal at elevated temperatures (from 1200°F, 650°C, for aluminum alloys to 3000°F, 1650°C, for steel) absorbs gases and non-metallic contaminants readily. Metal should be *poured smoothly* as a precaution against contamination, and should only be held in *clean, dry* containers. The distance

TABLE 12.2

	Bull Ladle	2-Man Ladle	Hand Shank	Lip-Pouring Ladle	Teapot-Pouring Ladle	Bottom-Pouring Ladle	Iron or Steel Ladle	Refractory-Lined Ladle
Magnesium	x	x	x	x			x	
Aluminum	x	x	x	x			x	x
Gray iron	x	x	x	x				x
Malleable iron	x	x	x	x				x
Ductile iron	x	x	x	x				x
Brass and bronze	x	x	x	x				x
Steel (basic)	x	x	x	x	x	x		x
Steel (acid)	x	x	x	x	x	x		x

molten metal travels through air during pouring should be shortened as much as possible, as the metal is particularly sensitive to gas absorption at this time. Streamlined flow is desirable at all times to the greatest extent possible. It is desirable to control pouring temperatures of molten metals as a precaution against excessive gas pickup. It is good practice never to disturb the surface layer of molten metal, as is usually done when ladles of non-ferrous metal are "skimmed" or "stirred" (unless scrupulously dry and clean tools are used, and an atmosphere of inert gas or slag is maintained over the bath).

12.13 Equipment for handling molten metal

Table 12.2 lists the ladle equipment used for handling molten metals for casting; Fig. 2.12 illustrates some of the various types of ladles. Iron or steel ladles are used for handling magnesium because the hot, molten metal reduces refractory oxides. Aluminum is seldom handled in teapot-pouring ladles because of its tendency to leave a skin of metal on the pouring spout which is difficult to clean from teapot-type ladles. Bottom-pouring ladles are only used for steels, where removal of slag from the pouring surface is often troublesome. Steel melted in a basic lined furnace is seldom poured from lip-type ladles as the slag is too viscous to control easily. Slag is never a problem when pouring brass, bronze, or aluminum, and so teapot- or bottom-pouring ladles are not required here.

12.14   General considerations of quality control

The raw materials, as well as the finished product, of a manufacturing concern, and even the equipment and methods of each plant should be subjected to some effective form of *quality control*. In the larger and more progressive foundries a quality-control department is maintained which is just as essential as the engineering, sales, or production department.   In one large captive foundry an 80-man department was maintained for quality control of the various phases of manufacture.   The application of statistical quality-control methods is gaining favor rapidly and is effectively reducing the numbers of men needed for this important task.

One of the important features of quality control is the requirement that management must know what needs to be controlled, how to control it, and the limits of quality required by each phase of the manufacturing process.   Foundries have left too much of this to chance in past years, and the economics of their process and the dependability of their product have suffered in corresponding measure.

The quality control of the finished casting is largely determined by the specifications imposed by the consumer.   For example, if X-ray and Magnaflux inspection, and a measurable degree of surface smoothness, etc., are required, these procedures must be carried out; a foundry should only agree to make work to such specifications if they are adequately equipped to conduct such tests effectively.   There is a regrettable tendency to cut corners in this regard and to ship castings which are of marginal quality.   The foundry should realize that high quality is just as important to its welfare in terms of future orders as it is to the consumer in terms of machining costs and delivery dates of machines for which good castings are critical component parts.   At any rate, this section will not discuss further the various methods for controlling the quality of the casting, since the specific tests are described in Chapter 9.   It will suffice to re-emphasize that quality control of the final product is essential and effective means for its conduct should be established.   This section will be concerned with quality control of raw materials and other important prerequisites to manufacturing.

12.15   Quality control in the foundry

Aside from effective organization and competent personnel to handle a quality-control program, a set of control charts, including statistical analysis of sampling frequency and methods (which varies in a calculable way with material and conditions of its use) must be determined. Once determined, these charts must be followed closely. All incoming raw materials should be considered as possible control items; some of them may be unimportant enough, in terms of economy and quality of final product, to be ignored, and sometimes suppliers' controls are rigorous enough to preclude more than a very infrequent check. Melting stock and scrap should be checked and sorted according to a routine specification, as should fluxes, coke, sands, clays, and binders. Sometimes a chemical check is adequate, but often a physical or performance acceptance test should be made. It is very easy to affect (or infect) hundreds of tons of castings by bad raw materials unless careful checks are made. This is particularly true in iron foundries where an unexpected carload of high-sulfur coke or a shipment of high-sulfur scrap might cause considerable loss of castings before being spotted. Besides being able to spot inferior raw materials before use, the foundry having a good quality-control system is less apt to receive subpar material from suppliers. An example of this is in the case of a foundry having excellent facilities for checking a certain commodity; the supplier, having shipped a carload of material, requested the foundry to please make their tests as soon as possible so that, in case the shipment was not acceptable, it could be rerouted to another foundry (presumably one with less good quality control).

One of the most critical materials in a steel foundry is the bentonite clay used for bonding sand; not only is a great deal of it used, but also bentonite is very prone to vary widely in quality from carload to carload. Since all the properties of a molding sand depend primarily on the quality of the binder, it is important to be able to appraise its performance before accepting shipment. Very recently a "liquid limit" test [7] has become an approved standard of acceptance by the steel foundry industry for bentonites.

Quality-control tests for incoming new sands should include a chemical analysis of SiO_2 content and a cumulative curve showing grain size and distribution—also, some measure of clay content for natural sands. The properties of mulled sands which should be checked statistically are green and dry strength, green and dry permeability, and moisture content. If the clay and sand are checked adequately, these tests serve mainly to insure that the muller operator is carrying out his predetermined routine of adding proper ingredients and adhering to the proper mulling cycle. One of the weakest chains in the quality-control link seems to be inability to establish and maintain routine at the muller. After spending considerable time and money to develop, set up, and check muller controls, the average foundry superintendent will put the least-paid, least-trained, and least-interested man in the shop in charge of this important task. Sand is a common denominator for all the molders and core makers in the shop; more of this is used than of any other raw material, and yet its quality is more often than not left to chance. There are many other tests that can be made to control mulled sand quality, such as flowability, hot strength and deformation, etc., but these are primarily useful for research and development work and need only be used as an occasional statistical routine test. Cores should be tested for compressive or shear strength, permeability and color as routine control checks; the oil or resin should be given an incoming service test and core ovens should be carefully controlled by thermocouples properly placed to be sure baking is uniform throughout the oven.

Although machines and plant equipment come under the maintenance department for performance control, one simple category of equipment seems never to come under any department and is not generally very well controlled; this is the flask equipment. Next to sand, flasks are common equipment for all molders and core makers. Flasks that are weak or worn, warped, of wrong size and/or shape, and blessed with loose pins are responsible for an astonishing amount of scrapped castings. Flask quality must always be controlled, preferably by routine checks made at reasonably frequent intervals.

Molten metal can be checked for quality by rapid chemical analysis at proper intervals and by certain quick tests at the furnace. For example, a fracture test can be used to check the carbon content of steel and to determine the presence or absence of gases in bronze; a more accurate and rapid test for carbon in steel is the magnetic or hardness characteristics of a small, quenched specimen. Slag tests are useful, particularly for steel making in open-hearth and electric furnaces. In these tests the color of the slag cake and the flow length in a special mold measure properties of the slag which reflect melt quality. Fluidity tests, in which metal from the furnace is poured or sucked by controlled vacuum into a flow channel of suitable size, are very useful; since temperature (superheat) is the most significant single variable influencing the ability of molten metal to fill a mold, the fluidity test is also an accurate indicator of temperature. The use of a simple, spiral test, molded in green sand or a core, poured by ladling a sample of metal from the furnace, is particularly useful in electric-furnace steel melting where temperature measurement is costly and inconvenient; the fluidity test is less needed, except as a research tool, for the lower-melting-point metals where pyrometry is less a problem.

One very important and simple quality-control tool for molten aluminum is the so-called "vacuum degasser." Hydrogen gas seriously impairs the strength and ductility of aluminum alloy castings, and it is very important to determine its presence in the melt. In this test a small sample of the melt is dipped into a dry metal or ceramic crucible and placed quickly under the bell jar of a vacuum pump. As the metal solidifies under reduced pressure, any gas present in the melt forms exaggerated cavities in the freezing sample which can be assessed in severity by cutting the sample and observing the cross section. If relatively large amounts of gas are present, the top surface of the freezing specimen will puff out, and gas can be seen to be escaping. For small gas contents the test can be made quantitative by making a quick density test of a special type specimen.

There are many other specific items of foundry operation and materials meriting quality control, but enough has been said to indicate the importance of developing specifications for performance and raw materials acceptance; it is equally important to keep and use quality-control records. Workmen perform most dependably under a system of checks and balances and, also, good records permit quick solutions to problems. It is desirable to go over quality-control records with foremen, just as it has proved valuable to have daily clinics in which bad castings are placed on a table and causes of these discussed with the workmen. The value of the quality-control program should also be recognized and implemented by occasional management conferences. Until quality control and the use of statistical sampling is better understood it is probably desirable to seek the counsel of experts before setting up an extensive program; such men are available, as there is increasing emphasis on quality control in all manufacturing processes.

12.16 Summary

All types of melting units are found in foundries. In general cast iron is melted mostly in cupolas, with the arc and open-hearth units increasing slowly in popularity; air furnaces, a special form of the open hearth, are used extensively in shops making malleable iron. Steel for castings is commonly melted in a three-phase electric-arc furnace of size suitable for the particular operation; open-hearth furnaces are used in large foundries and ingot shops, as are cupola-converter combinations. Aluminum and copper-base alloys are melted in various types of crucible furnaces, indirect-arc furnaces, and small open-hearth units; magnesium alloys are usually melted in iron crucibles set in oil- or gas-fired furnaces. The induction furnace can melt nearly all commercial metals and alloys. The choice of melting unit may be based on one or more of nine influences mentioned in the introduction to this chapter; the chief factor is generally economy or availability of fuel for the area.

The basic raw material for making steel is *scrap;* for cast iron, scrap and pig iron; and for non-ferrous metals, primary and secondary ingot. The scrap and smelting (ingot-making) industries are vitally important components of foundry operations.

In melting of non-ferrous metals, pig or scrap of the proper composition for casting is usually employed as melting stock; no attempt is made to remove impurities (except gases) during melting. Melting operations for ferrous metals, especially steel, are usually designed to purify (refine) the metal. Easily oxidizable impurities are removed partially or completely by oxidation and slag-metal reactions. In basic melting (melting operations involving a basic slag and basic refractories) phosphorus and sulfur can be removed by slag-metal reaction if melt oxidation is properly controlled. A variety of *inoculation treatments* are used in foundries to effect desirable alterations of casting microstructures; these treatments alter structure of castings without appreciably changing their chemistry. Aluminum and magnesium alloys are *grain-refined* with minute alloying additions. *Graphitization* is promoted in gray cast irons with similar additions. Other agents transform gray iron (of proper purity) into *ductile iron,* and still others *modify* aluminum-silicon alloys.

In the last part of this chapter, the importance of establishing quality-control procedures in all phases of the foundry operations was emphasized. Effective quality control requires that management know what needs to be controlled, how to control it, and the limits of quality required. Today *statistical quality control* is becoming an established branch of engineering; experts are available, trained in its application to foundry as well as to other manufacturing processes.

12.17 Problems

1. List six factors that must be considered in selecting the type of melting unit for a foundry.

2. List the following fuels according to calorific value: bituminous and anthracite coal, coke, oil, coke-oven gas, producer gas, and natural gas.

3. List six distinctly different alloys (ferrous and non-ferrous), and indicate the type of furnace you consider ideal as a melting unit.

4. What factors make the rocking-arc type of furnace desirable as a melting unit?

5. What are the purposes of slags?

6. What two factors are considered in designing hearths of open-hearth furnaces?

7. List six metals or alloys that may be melted in the open-hearth-type furnace.

8. What practical factor limits the type of metal that can be melted commercially in an induction furnace?

9. Discuss briefly two advantages of induction melting. Why are these of interest to the foundryman?

10. Make a simple sketch of a typical cupola. Label the principal sections of the furnace, and describe briefly (but accurately) its operation.

11. Build a cupola of your own with an inside working diameter of about 3 in. Write to American Foundrymen's Society, Golf and Wolf Roads, Des Plaines, Ill., for details. You will learn a lot and have fun besides.

12. What comprises a cupola charge? What single material is used in the greatest tonnage in a typical cupola operation?

13. How are the impurities removed in a Bessemer converter? Include chemical reactions.

14. Why is no exogenous fuel required in operating a Bessemer converter?

15. What materials are used in preparing primary ingot? secondary ingot? Is it possible to make high-quality castings from secondary ingots? Explain.

16. What are the advantages and disadvantages of the coreless induction furnace as compared to the cored type?

17. List the steps you would take in gas-melting aluminum to assure clean, gas-free metal.

18. What steps might you take to assure a low-sulfur cast iron? Could you desulfurize in an induction furnace by melting under a basic slag? Explain.

19. What is the optimum type of slag for desulfurization? For dephosphorization?

20. How would you assure a low-hydrogen content in arc-melted low-carbon steel? How would you accomplish this for induction melting of carbon-free stainless steel?

21. Explain the principle of grain refiners. Would you expect undercooling to occur in grain-refined alloys?

22. Why are graphitizers so important in melting high-strength irons?

23. Compare the advantages and disadvantages of bottom-pouring ladles.

24. Would you use a teapot ladle for aluminum casting? Why or why not?

25. List the variables in sand practice that you might subject to quality-control procedures.

26. List the variables in electric-arc melting of steel that you might subject to quality-control procedures.

27. One of the simplest and cheapest, and yet very effective, melting units for non-ferrous metals is the air-blown, coke-fired type described herein; yet this furnace is not widely advertised commercially. Unless your school has such a furnace, why not get a group together and build one for your laboratory and for thesis project?

28. Design a small "do-it-yourself" type foundry kit for molding, melting, and pouring aluminum-base alloys.

29. If your project time is enough, build one—sometimes marketable items get their start this way.

30. Some metals melt at temperatures even below that of boiling water; find out what some of these are, and see to what extent you can simulate typical foundry procedures in "table-top" experiments.

12.18　　Reference reading

1. *Metals Handbook,* A.S.M., Cleveland, 1948.

2. *The Principles of Basic Open Hearth Steelmaking,* A.I.M.M.E., New York, 1950.

3. Sisco, F. T., *The Manufacture of Electric Steel,* McGraw-Hill, New York, 1924.

4. Briggs, C. W., *The Metallurgy of Steel Castings,* McGraw-Hill, New York, 1946.

5. Murphy, A. J., *Nonferrous Foundry Metallurgy,* McGraw-Hill, New York, 1954.

6. *Statistical Quality Control for Foundries,* A.F.S., Des Plaines, Ill.

7. Navarro, J. M., and Taylor, H. F., "The Significance of Liquid Limit in Evaluating Steel Foundry Bentonites," *S.F.S. Research Report,* **40**.

CHAPTER 13

Foundry Refractories

13.1 Introduction

Refractories (heat-resistant materials) are the backbone of the foundry. Refractories form the crucibles and furnace bottoms which contain metal during melting; they comprise the furnace walls and roof which reduce heat losses and contain gases. Molten metal is carried in refractory ladles and poured into refractory molds. Refractories are the common denominator of all processes carried out at elevated temperatures.

Foundry refractories are usually, but not always, oxides of elements (MgO, SiO_2, Al_2O_3, and mixtures of these and others). Fifty years ago these materials were obtained from ore bodies of uncertain purity, and used essentially as mined. Little attention was paid to chemical or structural properties; the ores were simply prefired where necessary, ground, and molded to form. Common ores employed were fireclay, quartzite (ganister), dolomite and magnesite.

The above refractories are still the principal ones used, but demands of the steelmaker and foundryman for longer-life and higher-temperature service have led to new developments in materials and to considerably greater control in refractory manufacture. Far more important, the increasingly stringent requirements have helped stimulate the large amount of fundamental research conducted only in the last few decades. Out of this research has come the beginnings of an understanding of refractories on a microscale—of the fundamental chemical and physical forces which make materials good refractories. In addition, research has resulted in tremendous improvements in the serviceability of traditional refractory materials and has created a wide range of entirely new refractory materials (the best known of these being magnesia from sea water and zirconia).

It is our purpose here to present the nature of foundry refractories from an engineering viewpoint, and to discuss refractories as they are used today in the foundry industry.

13.2   Classification of refractories

One classification of refractories for foundry use is on a chemical basis: i.e., *acid, basic,* and *neutral* refractories. Acid refractories are oxides of *metalloids,* such as SiO_2 and P_2O_5, and basic refractories are metal oxides, such as MgO and CaO. In the foundry, the best-known acid refractory material is silica (SiO_2) and the best-known basic refractory material is magnesia (MgO).

Just as acids and bases react in wet chemistry to form salts, so acid and basic refractories react to form compounds:

$$MgO + SiO_2 = MgSiO_3$$

$$CaO + SiO_2 = CaSiO_3$$

In many cases, the compound formed is a relatively low-melting-point liquid, or *slag.* For instance, the melting point of MgO is $5070°F$ ($2800°C$), the melting point of SiO_2 is $3135°F$ ($1720°C$), but the compound formed, Mg_2SiO_3, is a liquid slag above $2840°F$ ($1560°C$). Because such slags tend to form, eroding and weakening a refractory structure, acid refractories are not used in contact with basic refractories at high temperatures. Instead, *neutral* refractories are used to separate acids and bases. Neutral refractories are materials that react sufficiently slowly so that they can be used at high temperature in contact with either acid or basic refractories. Chrome ore and carbon are the major neutral refractories used in practice. The distinction between neutral and basic refractories is not as sharp as it was once thought to be. Chrome, for example, may properly be considered to be a basic material (Table 13.1), although it fulfills the requirements of a neutral refractory in ordinary metallurgical applications. Also, many refractory bricks are now made of *mixtures* of both chrome ore and magnesia (MgO); their degree of "basicity" depends on the amount of MgO in the brick.

Slags are an essential feature in the melting and refining of many metals. These slags are either *acid* or *basic,* depending on the

TABLE 13.1

TYPICAL COMPOSITIONS OF SILICA AND BASIC BRICK *

Per Cent

Type of Brick	Silica (SiO_2)	Alumina (Al_2O_3)	Lime (CaO)	Magnesia (MgO)	Iron Oxide (Fe_2O_3)	Chromic Oxide (Cr_2O_3)	Other Oxides
Silica							
Conventional and superduty (overall range)	95–97	0.20–1.2	1.8–3.5	—	0.3–0.9	—	0.05–0.3
Basic							
Chrome	3–6	15–33	—	14–19	11–17	30–45	1–2
Forsterite	33–39	—	—	47–55	9–11	—	3–4
Magnesite	3–6	0.4–2.0	1–5	85–95+	0.5–4.0	—	0.5–1.0

* Courtesy Harbison-Walker Refractories Co.

relative content of acid and basic oxides. In cast iron and steel foundries, the distinction is most important. Acid slags can be used when sulfur and phosphorus elimination (from the molten metal) is not necessary, but, when these impurities must be removed, basic slags are essential. *Basic slags* react with *acid refractories* just as basic refractories do, eroding and weakening the furnace structure. Hence, basic slags are commonly used only with basic refractories, and acid slags with acid refractories. There are exceptions to this general rule, but, when prolonged contact between slag and refractory is encountered, as in open-hearth furnaces, both slag and refractory should be either acidic or basic.

13.3   Refractory raw materials

Most refractory materials are obtained from ore bodies of reasonably high purity, as generally no chemical purification is employed. Over half of all refractories used are obtained from ores of essentially pure silica (quartzite), or from ores that are predominantly compounds of silica and alumina (fireclays and similar clays). Formerly, magnesia (MgO) was obtained by *calcining* (heating) magnesite, $MgCO_3$. At present large tonnages of MgO are also obtained by chemical treatment of magnesium bearing brine. Chrome (Cr_2O_3) is obtained from chrome ore of widely varying purity, and used in the impure form.

Some refractories are now being produced chemically from essentially pure elements or compounds. Silicon carbide (SiC), for example, is produced by high-temperature reaction of silica with carbon. Various special refractories, or "super refractories" such as thoria, beryllia, etc., are produced for specific purposes by other special leaching, chemical, and thermal methods.

13.4   Forms of refractories and refractory materials

After mining or chemical production, and calcining if necessary, refractory raw materials are crushed, ground, and sized. They are

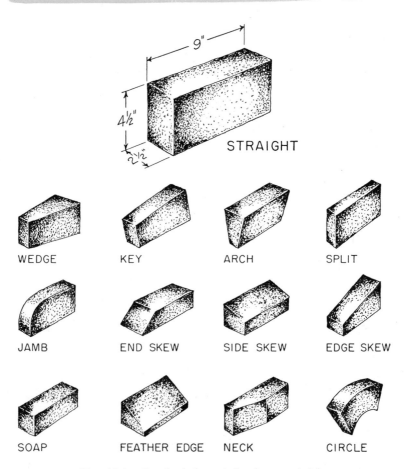

FIG. 13.1. Standard shapes of refractory bricks.

then mixed with other materials as desired and shipped to the consumer for use, or are fabricated into *shapes* such as bricks or blocks.

The foundryman uses refractories in many forms to suit his special needs; forms available include *prefabricated shapes, mortars, coatings, plastic mixes,* and *castables.* Use of these types of refractories in furnace construction is shown in the sketches of Figs. 13.8 through 13.12.

Most melting and other furnaces require brick (masonry) construction, and are lined or built up with prefabricated shapes. Such bricks are commercially available in a wide variety of sizes and shapes ranging from square blocks to keys, wedges, and arches (Fig. 13.1). Acid, basic, or neutral bricks of widely varying chemical and physical properties are readily obtainable. Some typical brick compositions are listed in Tables 13.1 and 13.2, and some characteristic properties of the bricks are tabulated in Table 13.3.

In brick construction a refractory *mortar* is generally used to bond the brickwork into a solid unit on firing, and to provide a cushion between the irregular surfaces of the bricks. One of the best mortars is a combination of a plastic clay with a volume con-

TABLE 13.2

TYPICAL COMPOSITIONS OF FIRECLAYS AND HIGH-ALUMINA BRICK *

	Per Cent			
Type of Brick	Silica (SiO_2)	Alumina (Al_2O_3)	Titania (TiO_2)	Other Oxides
Fireclay				
Superduty	49–53	40–44	2.0–2.5	3–4
High duty, aluminous	51–60	35–40	1.7–3.3	3–6
Semi-silica	72–80	18–24	1.0–1.5	1.5–2.5
Medium duty	57–70	25–36	1.3–2.1	4–7
Low duty	60–70	22–33	1.0–2.0	5–8
High alumina				
50% alumina class	41–47	47.5–52.5	2.0–2.8	3–4
60% alumina class	31–37	57.5–62.5	2.0–3.3	3–4
70% alumina class	20–26	67.5–72.5	3.0–4.0	3–4
80% alumina class	11–15	77.5–82.5	3.0–4.0	3–4
90% alumina class	8–9	89–91	0.4–0.8	1–2
99% alumina class	0.5–1.0	98–99	Trace	0.6

* Courtesy Harbison-Walker Refractories Co.

TABLE 13.3

CHARACTERISTIC PROPERTIES OF REFRACTORY BRICK *

Type	Class	Characteristics
High alumina	99% alumina 90% alumina 85% alumina 80% alumina 70% alumina 60% alumina 50% alumina	High refractoriness, which increases with increasing alumina content. High mechanical strength at high temperatures. Excellent to fair resistance to spalling. Greater resistance than fireclay brick to corrosion by most basic slags and fluxes.
Fireclay	Superduty	Stability of volume and high mechanical strength at high temperatures. Excellent resistance to thermal spalling. Fair resistance to highly acid slags; lower resistance to basic slags.
Fireclay	High duty Medium duty Low duty	The chemical and physical properties of fireclays vary between wide limits; hence, fireclay brick with widely varying combinations of properties are available. This fact accounts for their suitability for service under widely different operating conditions. Most fireclay bricks have relatively good spalling resistance and thermal insulation value. Fair resistance to acid slags and fluxes, lower resistance to basic slags and fluxes.
Fireclay	Semi-silica	Rigidity under load at high temperatures. Resistance to structural spalling. Volume stability. Resistance to penetration and attack by volatile alkalies or fumes. High temperature of incipient vitrification.
Silica	Superduty Conventional	High refractoriness and resistance to abrasion. High mechanical strength at high temperatures. Thermal conductivity at high temperatures appreciably greater than that of most high-duty fireclay brick. High resistance to corrosion by acid slags. Fair resistance to attack by oxides of lime, magnesia, and iron. Readily attacked by basic slags and fluorine. Not subject to thermal spalling at temperatures above 1200°F; poor resistance to thermal spalling at low temperatures.

TABLE 13.3 (*Continued*)

CHARACTERISTIC PROPERTIES OF REFRACTORY BRICK *

Type	Class	Characteristics
Magnesite	Fired 88–90% magnesia More than 90% magnesia Chemically bonded, metal-encased	Extremely high refractoriness and high thermal conductivity. Great resistance to corrosion by basic slags; poor resistance to slags containing high percentages of silica. The chemically bonded and metal encased brick have marked resistance to spalling.†
Chrome	Fired	High resistance to corrosion by basic and moderately acid slags and fluxes. In general, basic slags do not adhere to chrome brick. Under certain unusual conditions, iron oxide is absorbed and causes a damaging expansion. Thermal conductivity lower than that of magnesite brick but higher than that of fireclay brick.
Magnesite-chrome	Chemically bonded Chemically bonded, metal-encased Fired	Mechanical strength and stability of volume at high temperatures. Excellent resistance to spalling.† High resistance to corrosion by basic slags.
Chrome-magnesite	Chemically bonded Fired	Mechanical strength and stability of volume at high temperatures. Excellent to good resistance to spalling.† High resistance to corrosion by basic slags.
Forsterite	Fired	High refractoriness. Excellent strength at high temperatures. Marked resistance to corrosion by alkali compounds. Fair resistance to most basic slags; attacked by acid slags.

* Courtesy Harbison-Walker Refractories Co.

† Since basic refractories are less resistant to thermal shock than fireclay and high-alumina refractories, the terms used to express the relative spalling resistance of the basic group denote a comparison merely within that group, and not with other refractories.

stant grog (calcined clay), but many special compositions of mortars are available in which chrome, silicon carbide, silica, and alumina are used; the manufacturer of a brick should usually be consulted in selecting a mortar. After completion of the brick and

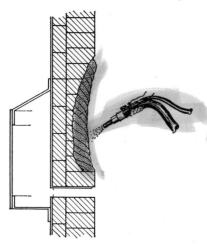

Fig. 13.2. Gunning a plastic refractory on a cupola wall.

mortar construction, the mortar is sometimes thinned with water and used as a *coating* for the face of the refractory walls to seal the joints further or to protect the wall from destructive elements in the furnace.

Plastic refractories (ramming mixes) are refractory materials that are *rammed* or *gunned* into place, as shown in Fig. 13.2. Plastic refractories are sometimes used instead of brick construction; an entire furnace wall may be rammed to produce a monolithic (one-piece) structure. Success with plastics in large monolithic structures depends on extremely skilled application, and so large furnaces are not generally constructed this way. Ladles, however, are often lined solely with a rammed mix. Perhaps the largest use of rammed mixes is in *patching* worn linings of cupolas and other furnaces.

A satisfactory acid plastic mix for many uses is fireclay and grog, with the percentage of grog high enough to prevent appreciable shrinkage during drying. Silica ramming mixtures (made from crushed quartzite bonded with clay) are also used. Basic ramming mixes usually consist of calcined magnesite with an organic bond such as tar or dextrine. Plastic mixes for furnace bottoms or other places where material will rest in place under the force of gravity require no room-temperature bond and are usually mixed only with water.

Refractory castables are mixtures of fireclay grog, high-alumina cement, and clay. They may be cast to shape. Sometimes rather large additions of organic matter are made to give the concrete better insulating characteristics. Concretes are only good for relatively low-temperature use and so are not widely employed in the foundry.

13.5 Structure of refractories

The structure of non-metallic refractories in many ways resembles the structure of metals. Refractories are generally crystalline (although rapid cooling may produce an amorphous *glass* structure). Pure refractory oxides and some compounds melt at a single temperature just like pure metals. Refractories combine with one another in the solid state to form *solid solutions, mechanical mixtures,* or *compounds.* Figure 13.3 shows the phase diagram for the important system SiO_2–Al_2O_3. The diagram shows the composition and temperature ranges (at equilibrium) for the various SiO_2–Al_2O_3 phases: cristobalite, mullite, and corundum. Figure 13.4 shows a rather different and simpler diagram for the CaO–MgO system. These compounds form a simple eutectic.

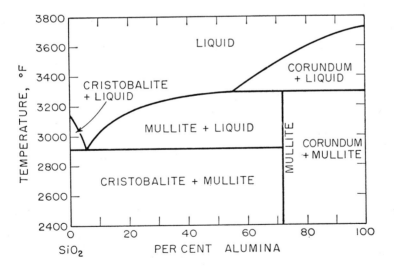

FIG. 13.3. Phase diagram for the silica-alumina system.

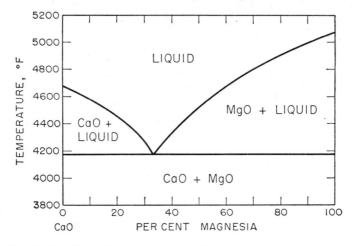

FIG. 13.4. Phase diagram for the calcium oxide-magnesia system.

The bonding between atoms in non-metallic crystals is very different from that of metallic crystals. In metals, at least the outer electron of each atom leaves its original mother atom and is free to wander throughout the lattice. The free electrons, or "electron cloud" of metals, account for many of the metallic properties as we know them: i.e., sheen, ductility, high electrical and thermal conductivity, etc. Refractories, on the other hand, have no free electrons. Their bond character is somewhat between *ionic* and *covalent,* and as a result refractories are generally brittle, possess relatively low electrical and thermal conductivity and medium to very high melting points.

A somewhat more practical difference between refractories and metals is that refractories are often quite serviceable mechanically when they are partially molten. Siliceous refractories may be capable of standing appreciable loads when they are as much as 30 per cent liquid. It is a general truth that most refractories at their higher temperatures of operation contain at least a small amount of liquid phase.

13.6   General considerations of acid refractories

Next to oxygen, silicon is the most abundant element in the earth's crust; together the two elements account for 75 per cent of all the materials readily available to man. Relatively high-grade deposits of silica (SiO_2) are available in many locations. Clay minerals, containing predominantly silicon, aluminum, and oxygen are even more common. Both these materials are easily molded to suitable shapes, and possess adequate refractoriness for most high-temperature uses. It is small wonder that silica and fireclay (acid) refractories are so widely used industrially.

Fireclay refractories, properly fired, expand and contract very little on heating and cooling, and are, therefore, very resistant to spalling (thermal cracking). Depending on their composition, they may possess adequate strength for most purposes up to about 2900°F (1590°C), and they are the least expensive of all common refractories. They are used for such purposes as cupola linings, heat exchangers for open hearths, insulating bricks, and for many other applications. Alone they account for over half the tonnage of refractories consumed.

The major advantage of silica over fireclay refractories is that it possesses superior strength as compared to fireclay at elevated temperatures (up to about 3000°F, 1650°C). It, therefore, finds extensive use in such applications as furnace roofs, especially open-hearth roofs. Well-made silica brick undergoes practically no thermal contraction or expansion at high temperatures, but even the best made brick undergoes *inversions* with corresponding volume change at lower temperatures (below 600°F, 315°C). Extreme care must be taken in heating such refractories in this temperature range. Repeated heating of silica bricks to high operating temperatures, as in a furnace roof, and cooling to room temperature causes excessive spalling and seriously shortens service life.

13.7 Fireclay and other alumina-silica refractories

Refractories that consist predominantly of silica (SiO_2) and alumina (Al_2O_3) are obtainable in a wide range of compositions, varying from essentially pure silica to essentially pure alumina (Fig. 12.3). Those containing approximately 20 to 40 per cent alumina are obtained from fireclays, kaolinite being the most important member of this group. Refractories containing 50 to 70 per cent alumina are usually fired diaspore clays, and refractories of still higher alumina contents are obtained from such ores as gibbsite, sillimanite, and bauxite. Table 13.2 lists typical compositions of bricks made from fireclay and high-alumina refractories.

In making fireclay bricks and other shapes, clays of various compositions are often mixed to obtain the proper composition on firing, and to facilitate the manufacturing operation itself. "Grog" (a previously fired clay material) is generally added in amounts over about 30 per cent to reduce shrinkage on firing. After grinding, and mixing, the raw materials are blended, *tempered* with water, molded to shape, and dried in air or low-temperature ovens. The bricks or other shapes are then *fired* at temperatures usually ranging from 2300 to 2600°F (1260 to 1430°C) for periods often as long as a week. The firing drives off chemically combined water, shrinks the bricks (to minimize volume change in service), and partially vitrifies the clays to obtain low porosity and high mechanical strength. The properties of a fireclay brick vary a great deal, depending on the thermal treatment, but, for metallurgical use, firing is usually at relatively high temperature for long periods to obtain dense bricks of high strength.

High-alumina refractories are more expensive than the fireclay refractories described. Alumina-silica refractories of about 70 to 80 per cent alumina are known as *mullites*. Bricks of this analysis develop the theoretical mullite composition after proper firing, as shown in the phase diagram of Fig. 13.3; they resist rapid heating and cooling cycles, and possess better high-temperature strength than ordinary fireclays. As the alumina content of brick is increased above the mullite composition, high-temperature strength continues to improve, and these bricks find use in such applications as roofs for indirect-arc furnaces where high refractoriness, good resistance to spalling, and high mechanical strength at high temperatures are essential.

13.8   Silica refractories

Refractories of essentially pure silica find widespread use, particularly for their load-bearing capacity at high temperatures. Like other refractories, they may be used in prefired brick form or rammed in situ. Standard silica brick contains approximately 95 per cent SiO_2, the remainder consisting chiefly of lime added to develop bond, and impurities of iron and aluminum oxide. The alumina is especially deleterious in reducing the fusion point of silica brick, as may be seen from the phase diagram of Fig. 13.3. A *superduty* silica brick is now obtainable which contains less than the normal amount of alumina and other impurities and, therefore, has a somewhat higher use limit than normal silica brick. Table 13.1 lists the composition range of silica bricks.

Silica may exist in at least six crystalline forms, not counting the vitreous or glassy state. The three major forms are *quartz, tridymite,* and *cristobalite*. *Quartz,* when heated for long periods at temperatures above 1600°F (870°C) slowly transforms to *tridymite* which in turn transforms to *crystobalite* at temperatures above 2680°F (1470°C). These reactions are very slow and are irreversible. At room temperature all three forms are quite stable. There exist, however, at least three other transformations which are *reversible,* and which proceed rapidly on heating. These include the transformation from low quartz to high quartz, and the other low to high transformations outlined in Fig. 13.5.

The transformation from low quartz to high quartz is particularly troublesome. It occurs with a substantial volume increase at 1063°F (573°C). To avoid the transformation and resultant spalling, silica bricks are fired to eliminate essentially all quartz and transform the structure to tridymite and cristobalite. Unfortunately these crystal structures also undergo reversible expansions (Fig. 13.5), but the expansions occur over a range of temperatures, and are less serious from a spalling standpoint. None-

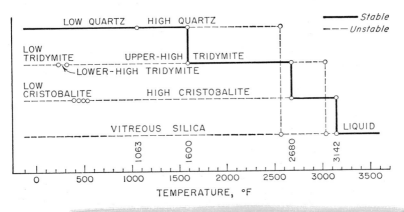

FIG. 13.5. Transformations and reversible behavior of pure silica (SiO_2).

theless, great care must be taken in heating or cooling these bricks at temperatures below about 600°F (315°C). At higher temperatures, reversible expansion of all forms of silica is practically nil—this being one of the great values of silica brick for the open-hearth roof, where the temperature fluctuates widely, but is never allowed to drop below 600°F, even for repairs.

13.9 General considerations of basic and neutral refractories

Basic refractories in general foundry use include *magnesite, magnesite-chrome,* and *burnt dolomite. Forsterite* is also used in certain applications, but usually not in direct contact with molten metals. Basic refractories in general use which have nearly neutral characteristics are *chrome-magnesite* and *chrome.*

Magnesite and magnesite-chrome brick are either *fired, chemically bonded,* or *metal-encased.* Fired bricks are heated to a high temperature in the manufacturing process to obtain a sintered bond; chemically bonded bricks are bonded with a low-temperature chemical bond, and sintering takes place in service. Metal-encased bricks are either fired or chemically bonded bricks which are partially contained in soft steel; in service the steel (essentially iron) is dissolved by the refractory and a monolithic structure formed. Burnt *dolomite* (Mg, Ca, O) was originally used only as a ramming mix for such applications as open-hearth furnace bottoms, since a simple, fired dolomite brick deteriorates quickly when exposed to the atmosphere at room temperature; the CaO in the brick reacts with moisture in the air. It has been found possible, however, to *stabilize* the calcium oxide by small additions of silica or iron oxide to the brick, and *stabilized dolomite* bricks are sometimes used in areas where there is a cost advantage over magnesite bricks. A basic brick known as *forsterite* brick, is of approximately the composition of the mineral forsterite ($2MgO \cdot SiO_2$). Its main advantage compared with other basic bricks is its relatively low thermal conductivity (Fig. 13.6). Table 13.1 lists typical compositions of basic bricks, and Table 13.3 lists some of their characteristic properties.

Compared with silica and fireclays, basic refractories all suffer the disadvantage of high thermal expansion. Figure 13.7 compares the expansion characteristics of several acid and basic bricks. The higher expansion coefficients (at furnace operating tempera-

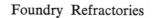

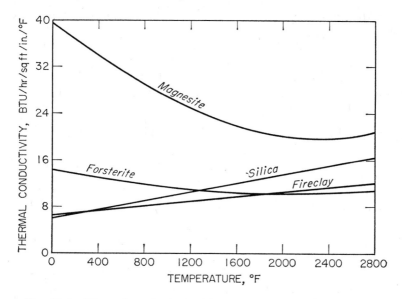

FIG. 13.6. Thermal conductivities of some standard refractory bricks.

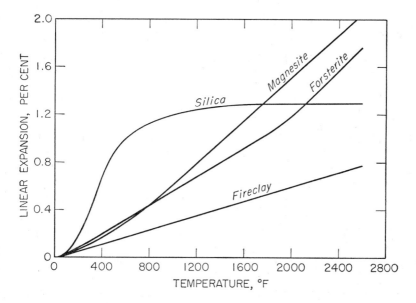

FIG. 13.7. Thermal expansions of some standard refractory bricks.

tures) of the basic bricks result in a greater tendency for these materials to crack and spall in service.

Since refractories which are strongly acid (such as silica) react
at high temperatures with strongly basic refractories (such as
magnesite), bricks of these materials cannot be laid in contact
with one another, for service at elevated temperatures. Instead,
bricks of a refractory with more nearly *neutral* characteristics must
be laid between the silica and magnesite. Bricks of *chrome ore*
have been most widely used for this purpose. The ore used is a
highly impure form of chrome oxide, containing only about 30
per cent Cr_2O_3. Originally chrome ore was used essentially as
mined, rammed in place with a suitable binder or fired into *chromite bricks*. These bricks possessed relatively low load-bearing
characteristics at the upper temperature of service, however, and
it was less than 20 years ago that the cause was laid to the high
iron content of the ore. Magnesia is now added to most chrome
ore to tie up the iron oxide. The "stabilized" chromite bricks are
known as *chrome-magnesite* bricks; they have come into widespread steel foundry use, both as a neutral "separating" refractory,
and also as a substitute for both acid and basic bricks in such
applications as the vertical walls of open-hearth furnaces.

Another refractory of essentially neutral characteristics is *carbon;*
it is used in foundry applications in either the ramming form (with
added bond) or as fired bricks. Carbon refractories enjoy extremely long life when conditions are not too oxidizing, and are
finding favor for use as linings in the basic cupola.

13.10   Refractories in the acid cupola

Refractory requirements in the cupola are among the most
severe encountered in melting furnaces. Even under the most
carefully controlled conditions, linings require some patching after
each heat. Failure and erosion of linings occur from many causes,
including abrasion from the charge, slag attack, mechanical strain,
excessive heat, and sudden temperature changes.

Figure 13.8 illustrates a typical arrangement of refractories in
a cupola. The furnace is essentially a vertical tube of refractory
bricks held in place by a steel shell. The most generally accepted
refractory for acid cupolas is high-duty fireclay (approximately 40
per cent Al_2O_3, 60 per cent SiO_2). The fireclay bricks used are
made with a particularly dense and uniform structure to resist

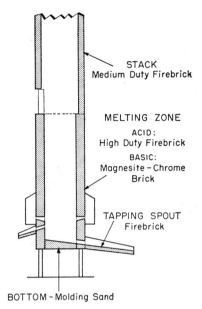

STACK
Medium Duty Firebrick

MELTING ZONE
ACID:
High Duty Firebrick
BASIC:
Magnesite – Chrome
Brick

TAPPING SPOUT
Firebrick

BOTTOM – Molding Sand

Fig. 13.8. Refractories in the acid and basic cupola.

abrasion and slagging. They possess relatively good thermal expansion characteristics (Fig. 13.7) and so are able to resist spalling, especially important when the blast of cool air enters the cupola at the end of a heat.

The most severe lining erosion in cupolas occurs in the melting zone. Here, not only is the temperature higher than in other zones of the cupola, but also chemical attack is most severe. Oxidation of the iron takes place in the melting zone and the oxides tend to react with the limestone flux and with the cupola lining. Abrasion occurs in this zone owing to the shifting of charge materials and concentration of blast air, especially as the refractory is weakened by the high temperature. Also, spalling is most likely to occur in and around the melting zone when the bottom is dropped (because of the inrush of cold air).

Most cupolas are lined with at least two courses of bricks, the total thickness of brick being from about 4½ to 9 in. One of the most common maintenance practices of such cupolas is to allow the inner course of bricks to erode at the melting zone until it needs replacing; then the entire course of bricks (in the region of the melting zone) is removed and replaced. For cupolas running 8 hours or more, such replacement is generally required after each heat. Another standard maintenance practice is to use a plastic mix, hand patching or gunning it in place (Fig. 13.2). Such ramming mixes, or "daubs," are usually silica with sufficient high-duty fireclay added to give good plasticity.

In the well of a cupola, service conditions are far less severe than in the melting zone, and large repairs are seldom necessary. The well, taphole, and spout are formed of fireclay brick, as is the melting zone, and the bottom is made of rammed molding sand. The bottom is, of course, replaced after each heat.

It is well to note that the service life of refractory materials in a cupola depends on the severity of the melting conditions, as well as the quality of the refractory used. The technique and skill with which the melting operations are performed affect the extent to which the cupola lining is damaged by each heat. No refractory has yet been developed that will withstand the service conditions of a cupola indefinitely; until one is found, care in selection and installation of refractories will be required, and careful control of the melting operation will remain a necessity.

13.11　Refractories in the basic cupola

Until recently, the relative simplicity and favorable economics of acid linings for cupolas resulted in their being used almost exclusively for melting iron. The development of ductile iron, however, brought with it a demand for base iron of low sulfur content, which is most dependably produced by a basic process. Many cupolas have been converted from acid to basic for the production of ductile iron, and many more will undoubtedly follow. As experience is gained in the use of basic linings for cupolas, their use may spread in some cases to the melting of gray iron for use as such. Hotter iron can be produced by the basic process, higher sulfur scrap can be used, and greater percentages of steel scrap can be melted successfully than is possible in the acid process. Initial installation costs of a basic lining are, however, considerably more than for an acid lining, and it is important to study fully the economics of basic cupola operation before conversion.

Linings for the melting zone of basic cupolas are usually of chemically bonded magnesite-chrome brick or high-fired magnesite brick. These basic bricks are laid from the bottom of the cupola up to about 5 ft above the tuyères (Fig. 13.8). The remainder of the cupola is lined exactly as the acid cupola, including the rammed bottom sand. In lining cupolas with basic brick, cognizance must be taken of (1) the high thermal expansion of basic bricks, and (2) their high thermal conductivity. "Expansion strips" of cardboard are usually placed between bricks in the melting zone to allow for the thermal expansion; the high thermal conductivity of the bricks requires that an outer course of fireclay brick be employed when the linings are less than about 9 in. in thickness.

Basic brick must be installed with basic mortars, and rammed patches must also be basic (or neutral). Plastic mixtures of carbon for ramming or gunning are widely used for patching basic cupolas because the carbon has a relatively long service life in the reducing atmosphere of the cupola melting zone. Recently, preformed carbon blocks have been used as lining for the melting zone in place of magnesite or magnesite-chrome bricks. One advantage of such a lining is that it can be used for either acid or basic melting; the slag composition can actually be changed *during* a melting operation without damage to the refractory. Large cupolas lined with carbon, as well as those lined with basic bricks, are sometimes *water-cooled* on their periphery during the melting operation to extend the service life of the refractory.

13.12   Open-hearth steel furnaces

As described earlier, in Chapter 12, open-hearth furnaces are used in a few large foundries for melting steel, and by nearly all shops producing ingots. Open-hearth furnaces are of either the *acid* or the *basic* type. The acid open-hearth is an efficient and economical furnace for melting large quantities of steel; however, it is not possible to use a basic slag with the acid lining, and so sulfur and phosphorus cannot be *refined* from the metal in the acid open hearth. The basic open-hearth process, on the other hand, employs a basic slag; it effectively removes phosphorus and decreases the sulfur content of the molten iron. Over 95 per cent of the total open-hearth production of steel in the United States is now of the basic type.

Figure 13.9 illustrates the types of refractories used in the construction of a basic open-hearth furnace. The working bottom

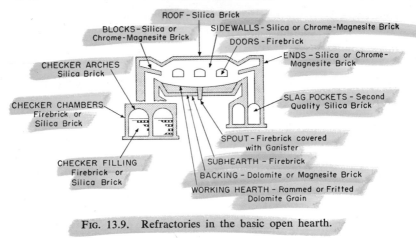

ROOF – Silica Brick
BLOCKS – Silica or Chrome-Magnesite Brick
SIDEWALLS – Silica or Chrome-Magnesite Brick
DOORS – Firebrick
ENDS – Silica or Chrome-Magnesite Brick
CHECKER ARCHES Silica Brick
SLAG POCKETS – Second Quality Silica Brick
CHECKER CHAMBERS Firebrick or Silica Brick
SPOUT – Firebrick covered with Ganister
CHECKER FILLING Firebrick or Silica Brick
SUBHEARTH – Firebrick
BACKING – Dolomite or Magnesite Brick
WORKING HEARTH – Rammed or Fritted Dolomite Grain

FIG. 13.9.   Refractories in the basic open hearth.

lining of such furnaces is generally of rammed or fritted dolomite; the bottom is repaired after each heat by throwing shovelfuls of loose refractory on the worn areas. Beneath the dolomite lining, magnesite or chrome bricks are used, and these in turn are backed with fireclay bricks for insulation. The front and back walls are usually made of chrome-magnesite brick, with magnesite bricks for the bottom courses. Silica was formerly used exclusively for the walls; in this case, a course of chrome or chrome-magnesite bricks was placed between the silica and magnesite bricks to prevent chemical attack.

The roof of the open-hearth is, from a refractory point of view, the most critical part of the structure; the life of the furnace as a whole is usually limited by roof life. Roofs fail because of the high temperatures inherent in steelmaking practice, and from attack by iron oxide and other "dust." Silica brick was formerly used almost exclusively for open-hearth roofs. Silica brick possesses superior structural strength at the high temperatures employed in the open hearth and is relatively inexpensive. Also important are the thermal expansion characteristics of this refractory. Operating temperatures in the open-hearth vary widely during the melting operation, but these fluctuations are well above the inversion temperatures of silica. At high temperatures (red heat) silica not only does not undergo rapid phase changes, but its thermal expansion is very low (Fig. 13.7). However, great care must be exercised in reheating if the roof temperature falls to a point where phase changes occur.

An increase of several hundred degrees in temperature in the basic open-hearth process should result in appreciable quality and economy improvement in the steelmaking process; however, silica roofs would collapse at such temperatures (about 3200°F to 3300°F). Basic bricks have been tried as a substitute for silica, to obtain a roof that will withstand higher temperatures, but the high thermal expansion, poor load-carrying ability, and tendency of such bricks to grow larger in use have slowed progress in this direction. However, in the last few years metal-encased chrome brick have been adopted in a great many open-hearth roofs with a high degree of success.

In *acid open-hearth* furnaces, silica refractories are employed almost exclusively above the charging floor level. The furnace bottoms are made up with mixtures of ground silica and fireclay, or simply with silica sand. Beneath this rammed or fritted layer, a subbottom is usually built of silica brick, and beneath this, one or more courses of fireclay and insulating firebrick may be laid.

13.13 Induction and direct-arc furnaces for ferrous melting

Tonnagewise, most steel is produced by the open-hearth furnace and most cast iron is produced from the cupola; both metals, however, are melted in substantial quantities in the *induction furnace,* and in the *direct-arc furnace.*

For ferrous melting in the induction furnace, linings are usually rammed in place (Fig. 13.10). The lining should be as thin as possible for highest melting efficiency. A satisfactory acid lining is rammed ganister (SiO_2) bonded with a little clay or sodium silicate. For a basic lining, dead burned magnesite with an organic binder is used; or the lining may be fritted in place with no bond by ramming behind a steel sleeve. When a steel sleeve is used, it is not removed; during the melting operation the magnesite sinters in place as the sleeve melts.

The direct-arc electric furnace is the most popular unit for melting steel for casting purposes. The refractory problems are much the same as those in the open-hearth process, and refractories used

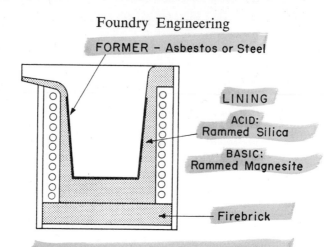

FIG. 13.10. Refractories in the induction furnace.

for both acid and basic melting are listed in Fig. 13.11. As in the open-hearth, one of the limiting features of furnace life is roof life. The problem is not as acute as that of the open hearth because in the electric furnace much of the heat from the arc is concentrated on the metal bath. On the other hand, electric furnaces are not always maintained at red heat; they are often shut down on evenings or weekends, and so *thermal spalling* is more of a problem. Silica bricks are usually used in spite of the spalling, but superduty fire-clay bricks also find wide application.

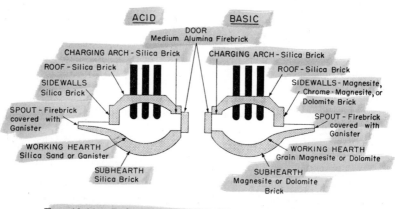

FIG. 13.11. Refractories in the direct-arc electric furnace.

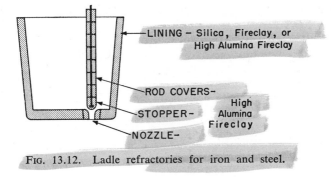

Fig. 13.12. Ladle refractories for iron and steel.

13.14 Ladle refractories for iron and steel

Figure 13.12 illustrates a typical bottom-pouring ladle and its refractories. The refractories listed are suitable for both iron and steel, and these refractories can be used for top-pouring ladles as well. It is interesting to note that high-alumina fireclay is chosen for the nozzle refractory because it erodes gradually. If the nozzle material did not erode at all, pouring rate would decrease with time as the level of metal in the ladle dropped. If the nozzle material eroded excessively, pouring rate would increase. By judicious choice of nozzle refractory composition, erosion can be made to just balance the loss of metal head, and a more or less constant pouring rate can be maintained throughout the tap.

Basic refractories are not normally used for ladles (even for pouring basic steel). Basic ladle refractories tend to crack easily, owing to their high thermal expansion, and tend to give a slag buildup which is difficult to remove without damaging the lining. Nonetheless basic ramming mixes are sometimes used as ladle materials, especially when ladle dephosphorization or desulfurization is practiced. Neutral ladle mixes, such as ramming mixes of carbon can also be used.

13.15 Refractories in the non-ferrous foundry

Most non-ferrous metals which are cast in appreciable quantities are less reactive than iron, or melt at a much lower temperature; hence, refractory problems in melting these alloys are not nearly as great as in the melting of steel. The ordinary cast non-ferrous alloys include alloys of nickel, copper, aluminum, magnesium, zinc, and others in lesser amounts.

Electric furnaces of many types are used in melting non-ferrous metals; there is no single "best" type. Induction furnaces and electrical-resistance furnaces are used, and may be lined with standard monolithic linings of either acid or basic materials, or prefired *crucibles* may be employed. In direct-arc melting of copper and nickel, furnaces similar to that of Fig. 13.11 are employed, with a magnesite hearth, magnesite sidewalls, and magnesite-chrome, silica, or alumina-silica roofs. In the indirect-arc furnaces ("Rocking Arc," Fig. 13.2a), high-alumina firebrick is often used. Service conditions on refractories in the indirect-arc furnace are the most severe encountered in the non-ferrous foundry; high-alumina firebrick, however, possesses adequate refractoriness since the rocking motion of the furnace swirls the metal over all the refractory, preventing localized overheating.

The gas- or oil-fired crucible furnace (Fig. 12.1) is the most popular of all furnaces for melting most non-ferrous metals. Prefired crucibles of materials such as fireclay, fireclay-graphite, magnesite, and silicon carbide are generally used. In the case of copper-base and aluminum-base alloys, fireclay-graphite crucibles have been used predominantly, although silicon carbide crucibles are growing in popularity. The silicon carbide crucibles have considerably longer life than clay-graphite, and their higher thermal conductivity sometimes permits more rapid melting; they are, of course, more expensive than clay-graphite.

Magnesium cannot be melted in ordinary refractory oxide materials; magnesium is such a strong oxide former that severe chemical erosion of the crucible occurs with resultant contamination of the metal. However, no search for special refractories has been warranted as crucibles of cast iron or stamped steel are of adequate refractoriness and give no troublesome contamination. Cast iron or steel crucibles are also widely used for melting zinc alloys, and, with proper refractory "washes," they may be used for melting aluminum.

In crucible-type furnaces (gas- or oil-fired, induction-, or resistance-heated), the crucible itself is often removed from the furnace after melting and used as a pouring ladle. When this is not done, the metal is transferred to a pouring ladle which is usually of rammed fireclay or prefired clay-graphite (for copper-base and aluminum-base alloys), and of iron or steel for alloys of magnesium or zinc. Sometimes light, stamped steel ladles are used in aluminum foundries for pouring small castings; these ladles are coated with a thin refractory "wash," generally a fireclay.

13.16 Molding materials

The mold into which hot molten metal is poured is a *refractory* in every sense of the word. Principles that apply in the selection of a refractory for melting apply also to the selection of a molding material; the major difference is that the *time* for reaction between molten metal and mold is far less than that for reaction between molten metal and crucible. Thus, for example, silica sand is an adequate molding material for magnesium alloys only because there is insufficient time for appreciable metal-refractory reaction to take place. Another difference between refractory requirements of furnaces and molding materials is that the molding material is subject to *thermal shock* which is far more severe than that imposed upon a furnace refractory, and provisions for this must be allowed. Silica sand is an adequate refractory for the casting of most foundry metals and alloys. For casting the more reactive metals, special refractories must be used, and these have been discussed in Chapter 2.

13.17 Summary

Refractories are common to all metallurgical processes conducted at high temperatures. In the foundry they form the crucibles and furnace linings for metal melting, and the ladles and molds for metal handling and solidification. Refractory materials are available in many forms to suit special needs; these include prefabricated shapes, mortars, coatings, plastic mixes, and castables. Most furnaces are built of prefabricated shapes (bricks), although some are lined solely with a rammed plastic mix.

One classification of refractories is on a chemical basis: i.e., acid, basic, and neutral. Acid refractories are made up largely of silica (SiO_2) and alumina (Al_2O_3); basic refractories are usually predominantly magnesia (MgO) or calcium oxide (CaO). Neutral refractories include chrome (Cr_2O_3) and carbon (C). At high temperatures, acid brick cannot be used in contact with basic brick, or in contact with slags; severe erosion of the lining would occur from chemical attack.

Fireclay refractories (silica-base refractories containing about 20 to 50 per cent alumina) alone account for over half the tonnage of refractories consumed. They possess adequate strength for most purposes up to about 2900°F (1590°C) and are the least expensive of all refractories; also, properly fired they expand and contract very little on heating and cooling. Fireclays therefore find extensive use for cupola linings, open-hearth heat exchangers, and for many other applications. Silica (nearly pure SiO_2) refractories are used for applications such as open-hearth roofs because of their superior load-bearing capacity (up to about 3000°F, 1650°C), and their very low thermal expansion coefficient at elevated temperatures.

For basic melting, a variety of basic refractories including magnesite, magnesite-chrome, and burnt dolomite are used. Chrome-magnesite, chrome, and carbon are used for their more neutral characteristics. Basic bricks are more expensive than acid, have a greater tendency to spall and crack in service, and are not as insulating as acid brick. In many cases, however, these disadvantages are counterbalanced by a longer service life of the basic brick.

Proper lining of metallurgical furnaces is essential to obtain optimum service life, melting efficiency, and metal quality. In the last part of this chapter, techniques and refractories used in lining a number of foundry furnaces have been discussed and illustrated.

13.18 Problems

1. What is meant by "acid refractories" and "basic refractories"? Why is this distinction important in steel melting?

2. What are the advantages of acid refractories for the open-hearth operation? What are the advantages of basic refractories?

3. Why were the roofs of open-hearth and electric-arc furnaces nearly always made of silica brick? What advantages are gained by using roofs of basic brick?

4. Compare the relative advantages and disadvantages of silica brick, fireclay brick, carbon brick, and magnesite brick for the lining of cupolas.

5. What are the differences between mortars, coatings, plastic mixes, and concretes?

6. Why are refractory bricks usually used to line large furnaces, rather than rammed mixes?

7. Why are basic refractories not generally used for ladle linings?

8. What are the advantages of silicon carbide crucibles as compared to clay-graphite for melting nonferrous metals?

9. What are the factors that affect service life of acid cupola refractories? Which ones are most important?

10. What are the factors that affect service life in the rocking indirect-arc furnace? Which are most important?

11. Obtain from a local foundry (a) a sample of synthetic molding sand and (b) a sample of natural bonded molding sand. Make a number of small cones from the sand, 2 in. in diameter by 1 in. high. Place the cones on graphite plates, and heat one cone of each sand to: 2500°F, 2700°F, 2900°F, 3000°F, 3050°F, 3100°F, and 3200°F. Hold at temperature for 15 minutes.

On the basis of your observations which sand would you recommend for casting steel?

12. Repeat Problem 11 (the first four temperatures), only place a piece of steel between the sand cone and the graphite.

Examine (with the aid of a microscope if possible) reaction products you obtain, and present your data and conclusions in a brief research report.

13. List three ways you could melt metal without the aid of refractories.

14. It is your job to develop a chloride base (NaCl, KCl, etc.) salt flux for use in an aluminum foundry. Examine applicable phase diagrams in the literature, and propose several compositions that would be suitable.

15. Obtain a sample of fireclay of known composition from a local firebrick manufacturer. Add a little water and ram into a miniature "brick" shape. Pick a series of firing temperatures on the basis of the fireclay composition, and with the aid of the phase diagram of Fig. 13.3 and what you know about brick manufacture.

Fire one brick at each temperature and examine for (a) shrinkage, (b) cracking, (c) sintering.

16. Repeat Problem 15, except prefire a substantial portion of your fireclay to make "grog."

13.19   Reference reading

1. Newell, W. C., *The Casting of Steel,* Pergamon Press, London and New York, 1955.

2. *Ceramics, a Symposium,* British Ceramic Society, Stoke-on-Trent, 1953.

3. Norton, F. H., *Elements of Ceramics,* Addison-Wesley, Cambridge, Mass., 1952.

4. Norton, F. H., *Refractories,* McGraw-Hill, New York, 1949.

5. *The Cupola and Its Operation,* A.F.S., Chicago, 1954.

6. *Basic Open Hearth Steelmaking,* A.I.M.M.E., New York, 1951.

7. Levin, E. M., McMurdie, H. F., and Hall, F. P., *Phase Diagrams for Ceramists,* American Ceramic Society, Columbus.

8. *Modern Refractory Practice,* Harbison-Walker Refractories Co., Pittsburgh, 1950.

9. *Refractories,* General Refractories Co., Philadelphia.

CHAPTER 14

Heat Treatment and Joining of Cast Metals

PART I. HEAT TREATMENT OF CAST METALS

14.1 Introduction

Heat treatment in many foundries was once relegated to a dark and smoky corner. Temperature control was poor or nonexistent, and furnace equipment inadequate. Often the only thermal treatment conducted was carried out in an oven which was also used for baking cores. Today, heat treating is a carefully controlled process. Modern demands for castings of consistently high mechanical properties have made heat treating a crucial step between the casting process and the engineering application.

Many different heat-treating processes are in commercial use today. Two of the most important are the *solutionize-quench-age* treatment for non-ferrous alloys and the *homogenize-quench-temper* treatment for ferrous alloys. Both these treatments can produce a material many times stronger than the "as-cast" alloy, and a material with substantially increased hardness and improved ductility and impact strength. Other heat treatments are used for various special purposes. *Stress relieving* is a heat treatment to remove internal stresses which otherwise might later cause warpage or failure. *Annealing, normalizing,* and *spheroidizing* are all treatments to produce special combinations of properties in steel castings. *Malleableizing* is the heat treatment whereby the cementite of white iron is transformed to graphite "nodules" of malleable iron, with consequently increased ductility. These and other important commercial heat treatments for cast alloys form the basis for the following discussion.

14.2   Stress relieving

When castings are shaken from the mold they often have surprisingly high internal stresses "locked" in the structure. The stresses arise from the uneven cooling rates of various sections of the casting and from resistance of the mold to normal solid contraction of the casting; these stresses can be large enough to result in severe warpage or actual failure of the part during subsequent finishing operations or in service.

To remove objectionable stresses, castings are often given a thermal *stress-relieving* treatment. This treatment may be the only heat treatment given a casting, or it may simply be preliminary to other operations. In steel castings, stress relieving is conducted by heating to about 1100°F (590°C) and furnace cooling. The treatment does not alter the metallurgical structure of the casting; it is only to permit the locked stresses to "creep" out. In other metals, stress relieving is generally conducted at somewhat lower temperatures; gray cast iron is heated to temperatures between 800°F (430°C) and 1100°F (590°C), and most nonferrous metals are heated to 400 to 500°F (200 to 260°C). Times required at the above temperatures are variable but are generally between 1 and 5 hours. These times are usually stipulated in pertinent specifications. Some quantitative measurements have been made and published on the rate of stress relief in cast iron and steel at various temperatures.[13]

A problem sometimes arises when previously quenched and tempered castings require stress relief. Often the temperature necessary for complete stress relief is well above that of final tempering temperature; the use of such temperatures would seriously alter the structure and mechanical properties of the casting. In such cases, a compromise between stress relief and properties is usually made.

Castings too large to be completely heat-treated can be given *local stress-relief* treatments. Areas known to be subject to internal stresses are simply heated with the aid of a gas torch or other device. The major disadvantage of this method is that it can cause severe stresses at other portions of the casting, in the region between the heated and unheated zones.

Stress relief can also be accomplished wholly or partly by mechanical means, although such methods are not nearly as common as thermal means. The surfaces of castings are sometimes *shot-peened* to eliminate surface tensile stresses, and introduce surface compressive stresses; the latter stresses are desirable in that they improve fatigue life. Residual casting stresses can also be redistributed by *stressing beyond the yield point*. One example of this is "overspeeding" turbine wheels and the use of *"autofrettage"* for prestressing large gun barrels. The excessive speed or the internal pressure in the autofrettage process induces some plastic flow in the metal, reduces undesirable internal stresses, and adds desirable stresses which act to counteract service loads. Stresses can be reduced in large, complicated castings by inducing vibration in them mechanically at low or room temperatures, but this is a relatively little-used method.

14.3   Solution treatment of non-ferrous alloys

All cast alloys are subject to microsegregation. The segregation may be only a concentration gradient within the dendrite arms, or it may be a second phase precipitated at the grain boundaries. Examples of these types of microsegregation were discussed in Chapter 4. *Solution (homogenization)* treatments are often used to minimize or eliminate the concentration differences. Such treatments (1) improve casting ductility, especially in ingots which are to be subsequently worked, or (2) dissolve a maximum amount of certain desirable constituents in the lattice structure to make the alloy amenable to subsequent hardening treatments.

An example of solution treatment for the latter purpose might be chosen from nearly any one of the higher-strength aluminum-, magnesium-, or copper-base alloys; typical of these is the aluminum–4.5 per cent copper alloy. As this alloy cools in a sand mold, solidification is too rapid for appreciable diffusion to occur in the solid. The result is *coring* and *eutectic segregation*. Further, on cooling below the solidification temperature the alloy enters a two-phase field of the phase diagram and additional beta constituent ($CuAl_2$) precipitates at the dendrite boundaries and within the dendrites (Fig. 14.1a). If the cast alloy is subsequently heated to a temperature of about 960°F (516°C), it is in a single-phase field (according to the phase diagram) and the beta phase will tend to dissolve. At this temperature, rates of diffusion are much faster than at room temperature; the rate of solution and the elimination of concentration gradients will occur in a reasonable length of time

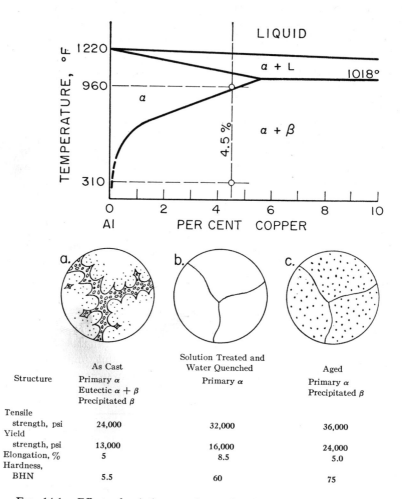

Structure	As Cast	Solution Treated and Water Quenched	Aged
	Primary α	Primary α	Primary α
	Eutectic $\alpha + \beta$		Precipitated β
	Precipitated β		
Tensile strength, psi	24,000	32,000	36,000
Yield strength, psi	13,000	16,000	24,000
Elongation, %	5	8.5	5.0
Hardness, BHN	5.5	60	75

FIG. 14.1. Effect of solution treating and aging on the structure and mechanical properties of an aluminum–4.5 per cent copper alloy.

Note: Precipitate in aged sample is usually too fine to see even under the microscope.

(8 to 16 hours). When the alloy is completely solutionized, it is *quenched* (cooled rapidly) to room temperature. By rapid quenching it is possible to retain the homogeneous, single-phase structure at room temperature (Fig. 14.1*b*), even though the phase diagram indicates that another phase should precipitate; this is because diffusion rates at room temperature are very slow. The aluminum-copper alloy must be quenched in water to assure that no precipitation occurs on cooling; other metals may be cooled more slowly and still obtain the desired result. In such cases, quenching in oil or even air is adequate.

Heat-treatment specifications and handbooks list the type of quench necessary for commercial foundry alloys.[1, 2] Most aluminum and magnesium castings for structural or load-bearing uses produced today are sold in the heat-treated condition, and most of these heat treatments involve a solutionizing step. Often the treatments are conducted at temperatures not far below the melting point of the alloy; great care must be taken to assure that even slightly excessive temperatures are not encountered in any portion of the heat-treating furnace. Care must also be taken to assure that temperatures are not much below those specified. Diffusion rates decrease very rapidly with decreasing temperature, and proper solutionizing often cannot be obtained when the furnace temperature drops by as little as 10 or 20°F. Specifications generally require that solution-treatment temperatures be held within ±10°F, and closer control than this is often desirable. Furnace control equipment used for solutionizing treatments should be checked for accuracy at frequent intervals and thermal surveys should be made of the furnace to determine uniformity of temperature throughout.

14.4   Age hardening

Solution treatments improve the ductility of cast non-ferrous alloys, but affect yield and ultimate tensile strengths only slightly. To improve these properties, *age hardening* is frequently employed. A solution-treated alloy, such as the aluminum–4.5 per cent copper alloy (Fig. 14.1) is a *supersaturated solid solution* after it is quenched to room temperature. The compound beta tends to precipitate in accordance with the phase diagram, but is prevented from doing so by the very slow diffusion rates involved. With long times, however, or at higher temperatures, the beta is able to precipitate in very fine particles, and these particles in the alloy matrix act to increase the tensile strength of the alloy, at the expense of some ductility (Fig. 14.1c). Figure 14.1 lists typical mechanical properties of the aluminum-copper alloy in the as-cast, solution-treated, and aged conditions.

Some aluminum alloys age sufficiently rapidly at room temperature that treatment at elevated temperature is unnecessary; others, such as the alloy of Fig. 14.1, require an aging treatment at 300 to 400°F (150 to 200°C) for 3 to 10 hours. Magnesium alloys are generally aged at about these same temperatures. Aging treatments for common non-ferrous foundry alloys are listed in pertinent specifications,[1] and elsewhere.[2] Temperature control in the aging cycle is equally as important as it is in the solutionizing treatment.

14.5   Time-temperature-transformation curves for steel castings

When carbon-steel castings are heated above a definite temperature, the metal structure becomes a single-phase solid solution, just as do many non-ferrous alloys. In the case of steel the solid solution is called *austenite*. The temperatures necessary to form an austenitic structure depend on the chemical analysis of the steel part; for plain carbon steel the temperature can be predicted from the iron-carbon phase diagram of Fig. 14.2 (shown in more detail in Fig. 4.11).

As a carbon-steel casting is cooled below the temperature where austenite is stable, transformation products begin to appear in the microstructure. These products are strongly dependent on the rate of cooling, as are the mechanical properties of the casting. Consider a eutectoid (0.80 per cent C) steel. If a small casting of this composition is furnace-cooled after being heated to the austenite temperature range, the structure transforms to coarse pearlite, as sketched in Fig. 14.2. If the cooling rate is somewhat more rapid, as in *air cooling,* the pearlite formed is finer, and the finer structure results in higher strengths. Still faster cooling results in an even finer structure with still higher strengths. Finally, a cooling rate may be reached which is so rapid that no pearlite is found in the microstructure; the structure is hard and strong, but is extremely brittle. This is *martensite* and is useful mainly in that subsequent tempering operations can be used to alter the martensite to a structure with excellent combinations of strength and ductility.

The concept of the *time-temperature-transformation* curve is useful in predicting the structure (and properties) that will be obtained on cooling steel from the austenitizing temperature. Time-temperature-transformation curves are also termed *TTT* curves, S curves, and C curves; a typical example is shown in Fig. 14.3.

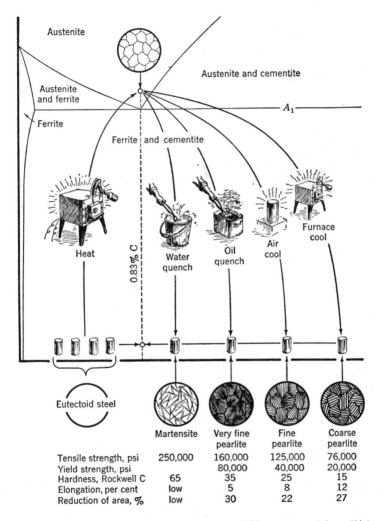

	Martensite	Very fine pearlite	Fine pearlite	Coarse pearlite
Tensile strength, psi	250,000	160,000	125,000	76,000
Yield strength, psi		80,000	40,000	20,000
Hardness, Rockwell C	65	35	25	15
Elongation, per cent	low	5	8	12
Reduction of area, %	low	30	22	27

FIG. 14.2.   The results of quenching a cast 0.80 per cent carbon (½-in.-diameter bar).

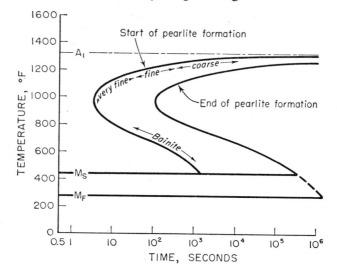

Fig. 14.3. Typical *TTT* diagram for an 0.80 per cent carbon steel.

Each point of a *TTT* curve is obtained by quickly cooling several small samples from the austenite region to a specific temperature below the region; individual samples are held at that temperature for differing lengths of time and then quenched in water. Metallographic examination is used to determine the time required at each temperature for transformation to begin, and the time required for transformation to end. For example, in determining the points at 1000°F (540°C) on the *TTT* curve of Fig. 14.3, samples would have been quickly cooled from above the A_1 temperature to 1000°F and held for different times, perhaps 1, 5, 10, 100, and 1000 seconds; at each time interval, one sample would have been quenched to room temperature. On metallographic examination, it would have been observed that the sample held 5 seconds was just beginning to transform to pearlite, and the sample held 100 seconds was essentially all transformed to pearlite; these two times are then the "start" and "end" times of transformation (at 1000°F) on the *TTT* curve. Similar points for a series of different temperatures, plotted together make up a curve such as that of Fig. 14.3.

Figure 14.3 illustrates the transformation characteristics of an 0.80 per cent C steel. When transformation takes place just below the "critical" temperature (below the A_1; 1333°F, 723°C), a relatively coarse pearlite is formed, and a relatively long time is required for transformation. At lower temperatures transformation is more rapid, and a finer pearlitic structure is obtained. In the 0.80 per cent carbon steel, most rapid transformation occurs at about 1000°F (540°C). This position of the TTT curve is termed the "nose"; the position of the nose is extremely important in heat-treating steel, and will be discussed further later in this chapter. At temperatures just below the nose of the curve, pearlite takes somewhat longer to form, and at still lower temperatures the transformation product changes character somewhat and is termed "bainite."

If the steel alloy is quickly cooled to temperatures below 280°F (140°C), and the microstructure examined, a surprising result will be found. The microstructure is entirely transformed, and this transformation is not dependent on time; it goes to completion almost instantaneously. The structure is *martensite;* it forms from austenite by a mechanical shearing action and so does not require time for diffusion as do most other transformations. At temperatures between 425°F (220°C) and 280°F (140°C) the structure is partially martensitic with the remainder untransformed austenite; these temperatures are shown as M_s and M_f, respectively (Fig. 14.3). The amount of martensite depends only on the temperature (increasing with decreasing temperature) and not on time.

As described above, TTT curves are obtained by rapid cooling of thin samples to a particular temperature, and holding the samples at that temperature to obtain isothermal transformation. The TTT curves do not strictly apply to the case of continuous cooling which actually occurs in the quenching operation. However, a *qualitative* picture of the structure to be expected in a given quenching operation may be obtained by superimposing the quench-cooling curves of a given casting on the TTT diagram. The comparison is especially valid if it is remembered that continuous cooling tends to shift the TTT curves downward and to the right. Consider, for example, a large cast steel bar (0.80 per cent C), quenched from above the A_1. The surface of the bar cools very rapidly and transforms entirely to martensite; the mid-radius cools less rapidly and transforms to fine pearlite; the center cools still more slowly and so transforms to a coarse

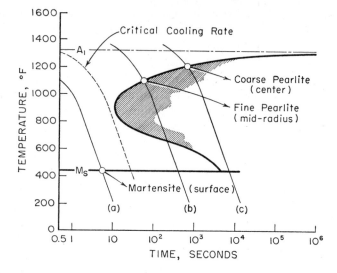

FIG. 14.4. Typical *TTT* diagram for an 0.80 per cent carbon steel. Cooling curves during quenching of a large cast steel bar are included; the cooling curves are shown for the surface, mid-radius and center of the bar.

pearlitic structure. Figure 14.4 shows these cooling curves super-imposed on the applicable *TTT* diagram.

TTT curves for a large number of commercial steels are listed in atlases.[3, 4] Although these curves have, for the most part, been determined for wrought steels, the data are applicable to cast steels of comparable compositions.[5] The actual shapes and positions of the curves are somewhat affected by structural variables, such as grain size and degree of homogenization, but the composition of the steel is most important in determining the *TTT* curve. Typical curves for steels of two different compositions are shown in Figs. 14.5 and 14.6. Figure 14.5 shows the *TTT* curve for a 0.4 per cent C steel. In this steel, at sufficiently slow cooling rates, ferrite tends to form before pearlite, as would be expected from the iron-iron carbide phase diagram. The *TTT* curve has, therefore, an added line denoting the beginning of ferrite formation. Note that, even in a steel of this low carbon content, cooling could be adjusted to yield a totally pearlitic structure. Figure 14.6 illustrates the *TTT* curve of a low-alloy steel containing 0.4 per cent carbon; the "nose" of the curve is moved to the right

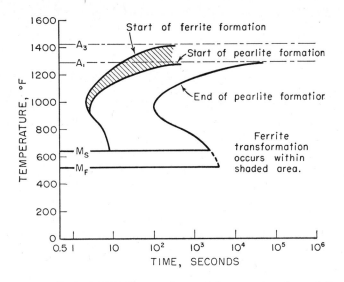

FIG. 14.5.   Typical *TTT* diagram for an 0.4 per cent carbon, plain carbon, steel.

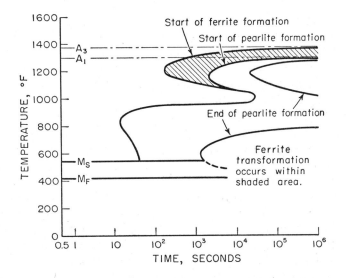

FIG. 14.6.   Typical *TTT* curve for an 0.4 per cent carbon, low-alloy, steel.

and the temperature of formation of martensite is decreased (compared to a plain carbon steel of the same carbon content).

14.6 Annealing and normalizing

Annealing (full annealing) is a heat treatment for steel castings which homogenizes the structure and refines the grain. Annealing produces a structure with good ductility and lower tensile and yield strengths than are obtained by other heat treatments. It is often used to improve machinability.

Annealing is performed by heating to the austenite region and holding long enough to allow homogenization of the austenite grains to take place; the casting is then *furnace-cooled* to allow formation of relatively coarse pearlite. Homogenizing temperature is usually 100 to 200°F (85–110°C) above the upper critical temperature (line A_3, Fig. 4.11); castings are held at temperature for 1 hour per inch of heaviest section.

A widely used heat treatment for steel castings is normalizing. In this treatment the homogenization is followed by *air cooling*. It results in improved tensile, yield, and impact properties, compared with the annealed material. In carbon and low-alloy steels the microstructure after normalizing is composed of pearlite with some ferrite. In more highly alloyed steels some bainite or martensite may be present; in the latter case normalizing is followed by a tempering treatment. The structure obtained by normalizing is quite sensitive to variations in cooling rate; mechanical properties vary substantially with varying section size. Care must be taken to assure that cooling is as uniform as possible throughout; for example, it is poor practice to stack castings too closely together, or to leave them on a hot truck hearth during the cooling cycle.

14.7 Quenching steel castings

The heat-treating cycle which (when feasible) produces the optimum mechanical properties in steel castings is the quench and temper treatment. This treatment involves *quenching* from austenite temperatures at such a rate that the only transformation product formed is *martensite*. A subsequent *tempering* operation is carried out primarily to improve ductility.

To form a completely martensitic structure, the quenching must be sufficiently rapid that cooling rates in all portions of the casting are greater than the "critical cooling rate" shown in Fig. 14.4. In most cases this involves quenching in either an oil or a water bath, water being the more rapid of the two. Many steel alloys have the "nose" of the *TTT* curve displaced so far to the left that it is difficult (or impossible) to cool sufficiently rapidly to avoid forming pearlite. For example, all portions of an 0.80 per cent C steel casting would have to cool to below 1000°F in less than 5 seconds in order to completely transform to martensite (Fig. 14.3). Such a cooling rate is impossible, regardless of quenching media, for the center of a heavy steel casting.

In order to make it possible to obtain martensite more easily, a number of alloying elements are added to steel: most notably, nickel, chromium and molybdenum. These elements move the *TTT* curve far to the right and permit martensite to be formed at much lower cooling rates; in the language of the heat treater, they increase the *hardenability* of the steel. Steels that contain alloying elements primarily for improved hardenability are termed *low-alloy steels*.

The quenching operation is probably the most delicate of all steel heat-treating operations. Heavy castings of uniform section can be quenched in water with little difficulty. A variation in section size, however, results in uneven cooling and consequent internal stresses. These stresses are frequently enough to crack or distort the casting, and a great deal of ingenuity is often required to circumvent the problem. Oil or an air blast should be used as a quenching medium when hardening is possible by this method. Steel parts can be quenched in *jigs* to minimize distortion. Thin sections can be removed from the quench tank before thick ones, or the entire casting can be removed (and quickly tempered) before it reaches room temperature. In spite of these and other precautions, intricate castings are sometimes cracked in the quenching operation.

14.8   Tempering

A *tempering* treatment always follows quenching, and may follow normalizing. Tempering is carried out at temperatures varying from slightly above room temperature up to about 1300°F (700°C). Tempering times vary up to about 1 hour at temperature for each inch of heaviest section. The essential function of the tempering treatment is to impart some ductility (and impact

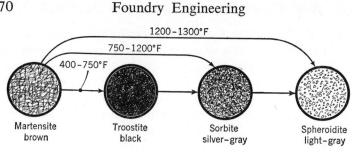

FIG. 14.7.   Tempering of martensite.

strength) to the alloy.   This is done by bringing about precipitation
of carbides so that, after tempering, the structure of a previously
martensitic casting becomes ferrite with precipitated carbides.   The
size of the carbides varies with tempering temperature and increases
with increasing temperature.   The fine structure obtained at low
tempering temperatures is termed *troostite;* at higher temperatures,
sorbite; and still higher, *spheroidite* (Fig. 14.7).

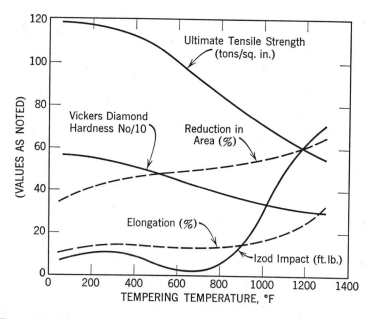

FIG. 14.8.   Properties of a quenched and hardened nickel-chromium steel.

The mechanical properties of the steel casting are markedly dependent on tempering temperatures.   Figure 14.8 illustrates how these vary for a nickel-chromium steel.   Note the low impact strength obtained at tempering temperatures between about 500 and 750°F (260 and 400°C); tempering temperatures in the region where this brittleness occurs should be avoided.   In some steels, tempering in the range of 800 to 1100°F (425 to 600°C) followed by slow cooling can also induce a form of temper brittleness.   This variety is *not* found in plain-carbon steels; it has been reported only in steels containing more than 0.70 per cent manganese or appreciable amounts of nickel and chromium.

14.9   Heat treatment of gray iron

For most purposes, gray iron castings do not require heat treatment.   By adjusting alloy composition it is usually possible to produce a casting with requisite strength and hardness in the as-cast condition.   Stress-relieving treatments, except for parts to be machined to close tolerances and for parts of widely varying section size are not generally required by specifications.

A variety of thermal treatments are available, however, for special purposes.   Stress-relief treatments have been mentioned earlier; these are usually conducted at 800 to 1100°F (430 to 600°C) for gray iron.   Sometimes it is advisable to improve the machinability of gray iron castings by an annealing treatment; the treatment chosen may be a low-temperature anneal, medium anneal, or high-temperature anneal.   Low-temperature anneals are designed to convert any pearlite present to graphite; they are performed by heating the casting to just below the critical (1300 to 1400°F, 700 to 760°C) and furnace cooling or air cooling.   Medium anneals are carried out at higher temperatures, 1450 to 1650°F (790 to 900°C), and are used when sufficient pearlite solution is not obtained at lower temperatures.   High-temperature anneals are similar treatments at 1650 to 1750°F (900 to 950°C); they are used if massive carbides are present in the cast structure which must be dissolved.   Figure 14.9 illustrates these annealing cycles and the structures obtained.

If gray cast iron is heated to just above the critical temperature region (1333°F, 723°C, for pure iron-carbon, 1350 to 1500°F, 730 to 820°C, for commercial gray cast irons), the flake graphite

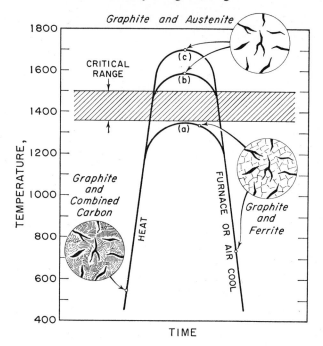

FIG. 14.9.   Annealing treatments for gray iron.   (*a*) Low-temperature anneal.   (*b*) Medium-temperature anneal.   (*c*) High-temperature anneal.

is left essentially undissolved but the matrix becomes an austenitic iron, of about 0.80 per cent carbon. Thus, the matrix is essentially an 0.80 carbon steel and can be hardened in the same manner as steel. Heat treatment does not greatly improve the mechanical properties of most gray irons, since, no matter how strong the matrix may be, the flake graphite particles act as stress raisers to cause failure at relatively low loads. Heat treatments, especially quenching and tempering treatments do, however, improve hardness and wear resistance and are sometimes used for this purpose. *Surface-hardening* treatments are quite often used. Here the surface is hardened by heating and rapidly cooling (and subsequent tempering), while the center of the casting is unheated and retains essentially its as-cast structure. Rapid surface heating is obtained by a gas torch (*flame hardening*), or by induction heating (*induction hardening*).

14.10   Heat-treating malleable iron

Malleable iron has been briefly discussed in Chapter 4. It is an iron of strength and toughness exceeding that of gray iron; the improved properties arise from the fact that the graphite in malleable iron is nodular in shape as compared to the flakes of gray iron.

Production of malleable iron begins with the casting of an iron of such analysis that it freezes *white* (has no free graphite). The casting is then heat-treated to obtain the desired structure and properties. The most common malleableizing treatment is a two-stage anneal to procure a totally ferritic matrix. In the first stage, the castings are heated to 1650 to 1750°F (900 to 950°C). At this temperature the massive carbides dissolve and "temper carbon" nodules precipitate. Castings are held at temperature until the structure consists of austenite and temper carbon. They are then cooled as rapidly as convenient to about 1400°F (760°C). The second stage of the anneal consists of slowly cooling through the critical region (1300 to 1400°F, 700 to 760°C, for ordinary malleable irons). The slow cooling through the critical dissolves any pearlite that may precipitate and reprecipitates it on the already existing graphite nodules. A complete annealing cycle requires from 18 to 72 hours, depending on section thickness and composition. The metallurgical structures obtained at various stages of the annealing cycle are shown in Fig. 14.10.

With today's increasing need for higher strength castings, the class of *pearlitic malleable* castings has been growing in popularity. The term is somewhat of a misnomer. It refers to a malleable iron which contains either pearlite or other iron carbide temper products in the matrix. Any of a number of heat-treatment cycles can be used in the production of "pearlitic" malleable iron. In one such cycle, the castings are air-cooled from a temperature somewhat below that of the first-stage anneal (but above the critical). This treatment results in a pearlitic matrix. Mechanical properties of malleable iron range from typically 60,000 psi ultimate tensile strength, 40,000 psi yield strength, and 20 per cent elongation for *ferritic malleable* to properties in excess of 100,000 psi ultimate tensile strength, 80,000 psi yield strength, and 2 per cent elongation for *pearlitic malleable* iron.

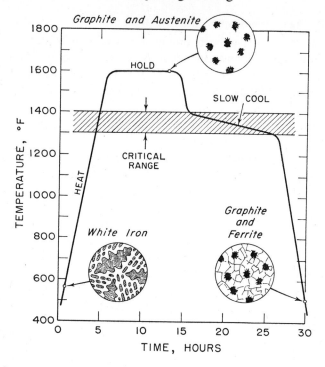

Fɪɢ. 14.10.   Typical annealing cycle for ferritic malleable iron.

14.11   Heat-treating ductile cast iron

Many ductile iron castings are used in the as-cast condition.
By proper control of composition and cooling rate castings can
be produced with an all-ferritic matrix (*ferritic ductile iron*) or
with an essentially all-pearlitic structure (*pearlitic ductile iron*) in
the as-cast condition.   Heat treatments may also be used to develop these structures.   If carbides are present in the as-cast structure, *ferritic ductile iron* can be produced by an annealing treatment at 1300 to 1400°F (700 to 760°C).   A *pearlitic matrix* can
be assured by normalizing.   *Martensitic ductile iron,* an iron with
a tempered martensite matrix, is obtained by quenching and
tempering.   Figure 14.11 illustrates the microstructures described
above, together with typical mechanical properties.

14.12   Equipment for heat treatment

Many different types of heat-treating equipment are used for many different purposes, and it is possible to describe here only a few of those that are widely used in production shops. The actual choice of furnace for a given heat-treating operation depends on many variables; the individual equipment makers should be consulted before one reaches a decision as to which type furnace is best suited to a particular operation.

In heat-treating non-ferrous alloys (other than stress relieving), extremely close temperature control must be maintained. Most

| Structure | Thermal Treatment | Typical Mechanical Properties | | | |
		Tensile Strength, psi	Yield Strength, psi	Elonga- tion, %	BHN
Graphite + ferrite	As cast (or annealed)	70,000	50,000	20	170
Graphite + pearlite	Normalized and tempered (or as cast)	110,000	80,000	6	270
Graphite + tempered martensite	Quenched and tempered	140,000	110,000	5	310

FIG. 14.11.   Structures and typical properties of as-cast and heat-treated ductile iron.

specifications for solutionizing and aging require temperature control to be better than $\pm 10°F$. Salt-bath furnaces are sometimes employed to achieve this uniformity, but recirculating air furnaces are generally more adaptable to production foundry operations. Figure 14.12 illustrates a furnace of this type. A powerful fan within the furnace maintains a rapid air circulation, minimizing dead "spots." A high-quality temperature controller is required to maintain accurate and constant temperature in this type of furnace. In the furnace shown, charging is generally accomplished by loading baskets outside the furnace; the baskets are then placed in the furnace by overhead crane. Care must be taken not to overcrowd the baskets or the air circulation will be impeded.

Homogenizing and other high-temperature treatments for steel do not require the extremely close temperature control of the recirculating air furnace. A furnace of the type shown in Fig. 14.13 is generally satisfactory. It consists simply of a furnace chamber, which is heated with gas, oil, electricity, or other fuel. In the furnace shown, charging is accomplished by loading a "car" outside the furnace and rolling the charge in on rails. Other furnaces are charged by lifting off the entire furnace roof with the shop crane. Smaller furnaces are charged by hand through a small door. Tempering of steel castings may be done in this type furnace, or in the recirculating air type, depending on the degree of

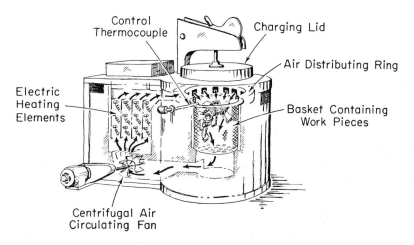

Fig. 14.12.   Recirculating air furnace.

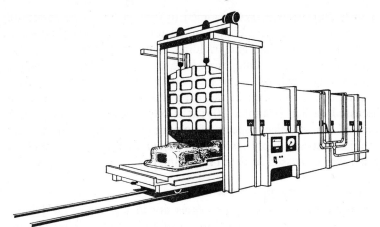

FIG. 14.13.   Car furnace.

control required of temper properties, and on the tempering temperature.

Steel castings may or may not be heat-treated with a controlled atmosphere in the furnace.   In the malleableizing of white iron, however, some provision must be made to prevent decarburization.   This can be done by either bedding the castings in a packing material or by maintaining a reducing atmosphere in the furnace during heat treatment.   Cast irons are usually treated in the same types of furnaces used for steel.

PART II.   JOINING OF CAST METALS

14.13   Introduction to joining

Some castings comprise a useful engineering or decorative part without being *joined* to other members.   Most castings, however, are joined to other parts before being used.   Methods for joining castings to castings or castings to other fabricated parts include bolting, riveting, resin bonding, and welding.   With the exception of welding, most of these joining techniques lie outside the province of the foundry.   Welding has always been a necessary adjunct to the casting process; it is used extensively for repair work as well as for producing complex engineering parts by joining several less complex components after casting.   Relatively simple cast shapes can be built into a complex *cast-weld* structure, or castings and forgings can be joined to form a *composite* structure. Care must be taken, of course, to assure a proper economic balance between casting and welding, but, when this balance is reached, inherent advantages of both the casting and the welding processes are realized. The advisability of considering cast-weld or composite structures in the early stage of designing has been discussed earlier (Chapter 10); it is our purpose here to present some of the technical aspects of joining by welding. These same technical aspects apply also to the *repair* of castings by welding, which was discussed in detail in Chapter 9.

14.14   Welding processes

Many different welding processes are in use today for joining metal parts. Some of these processes produce welds by the application of heat and pressure, but with little or no melting at the joint. For example, the blacksmith has for many generations forge-welded by hammering together hot iron parts. Modern welding techniques, such as butt welding and spot welding are based

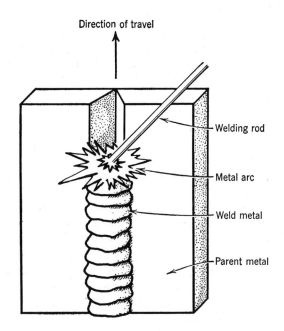

Direction of travel

Welding rod

Metal arc

Weld metal

Parent metal

Fig. 14.14.   Sketch of a weld in progress.

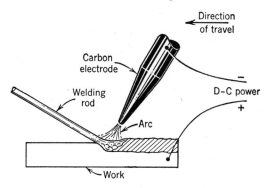

Fɪɢ. 14.15.   Carbon-arc welding.

on the same principle, the application of heat and pressure.   In these *non-fusion* welding processes, the source of heat usually used is the resistance of the part to the flow of electric current.

In the foundry, most joining and repairing are done by *fusion welding*.   Fusion welding involves actual melting at the weld area of the parts to be joined.   Additional molten metal is usually added to the weld area from a *welding rod*.   Fusion welding possesses many similarities to the casting process; it is, in fact, a process of continuous melting and resolidification throughout the length of a weld "pass," as illustrated in Fig. 14.14.   Figures 14.15 through 14.20 illustrate the welding processes normally used for cast metals.   Discussion of these processes comprises the remainder of this chapter.

14.15   Carbon-arc welding

In arc welding, the surfaces to be joined are fused by the heat of an electric arc.   The simplest type of arc welding, *carbon-arc welding,* is illustrated in Fig. 14.15.   In this process, the electric arc is drawn between the carbon rod and the work.   A separate *filler rod* may be fed into the arc at the desired rate.   Often fluxes, in the form of treated pastes or powders, are used as an electrode coating to shield the arc; the process is then termed *shielded carbon-arc welding*.   Carbon-arc welding is suitable for use with all ordinary foundry alloys, except those of magnesium.   It is used extensively (with suitable shielding) for welding copper-base and aluminum alloys.

14.16 Metal-arc welding

Metal-arc welding is a type wherein a metal electrode serves the dual purpose of carrying the arc and supplying the filler rod. Materials that can be welded by this process include most of the foundry metals (except magnesium). Great care must be taken in choosing a welding rod for joining a given metal; often the most desirable welding rod is not of the same composition as the base metal.

The electrode for metal-arc welding may be a bare wire, as shown in Fig. 14.16, or it may be *coated* with various materials (Fig. 14.17). Metal-arc welding with coated electrodes accounts for perhaps 80 per cent of all welding construction performed; nearly all welding of steel castings is performed in this manner. Coated electrodes produce denser, stronger, tougher, and more ductile welds than can be produced with uncoated rods.

Coatings fulfill a variety of functions. Those containing cellu-lose-type materials decompose slowly in the arc to produce a gas shield about the weld, and protect the molten metal of the weld from the bad effects of exposure to air. The reaction is $(C_6H_{10}O_5)_n + n/2O_2 \rightarrow 6nCO + 5nH_2$. Coatings containing *arc stabilizers* promote better ionization in the arc, causing it to be more stable. The melting rate is increased, as well as the depth of penetration of the weld. Because of the stabilizing effect on the arc, coated electrodes may be used with reversed polarity (work negative and electrode positive) in the welding circuit, thus allowing use of

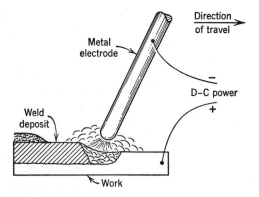

Fig. 14.16. Metal-arc welding with bare electrode.

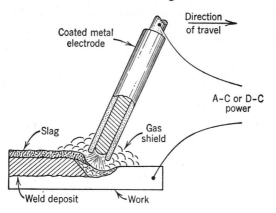

FIG. 14.17.   Shielded arc welding with coated electrode.

alternating current.   The main advantage of alternating current is that it is not subject to troublesome magnetic arc blow prevalent in d-c operation.   Titanium oxide and potassium compounds are considered arc stabilizers.

Welding rod coatings often contain *slag-forming* ingredients, deoxidizing elements, and alloying agents, in addition to arc stabilizers and cellulose.   *Slag formers* such as titanium oxide, calcium fluoride, and silicon dioxide protect the metal by forming a molten slag cover as the electrode melts.   This cover protects the metal deposit from the air and acts as a heat insulator to reduce the rate of cooling.   It also permits gases to escape, and reduces cooling stresses.   *Deoxidizers* such as ferromanganese react with oxygen from the weld melt, making denser and stronger welds.   *Alloying elements* like ferromolybdenum, ferrochromium, and elemental metals, carried in the coatings, mix with the weld metal during welding to give added strength to the joint.

An effort to obtain even better protection of the weld metal than that afforded by coated electrodes has resulted in the *submerged melt* process.   An example of this, the so-called *Union-melt* process, is illustrated in Fig. 14.18.   Here all melting and solidification takes place under a protective liquid slag.   This method is used extensively for joining pressure vessels and other large ferrous castings but has also been successfully used for welding many copper-base alloys.

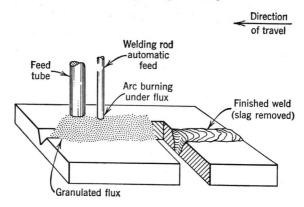

FIG. 14.18.　Submerged melt welding.

14.17　Inert-gas-shielded arc welding

Inert-gas-shielded arc welding is essentially a variant of the carbon-arc process with the addition of inert gas protection. Generally a tungsten electrode is used in place of carbon because of its longer life.　The process is illustrated in Fig. 14.19.　It was designed originally for welding magnesium and magnesium alloys.

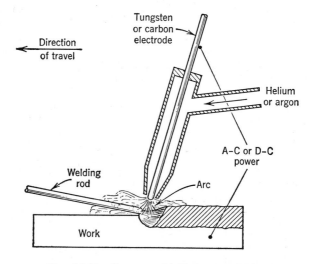

FIG. 14.19.　Inert-gas-shielded arc welding.

These metals are highly oxidizable and require protection from the atmosphere during welding. Today the process is also finding widespread use for welding aluminum. The method is more costly than metal or carbon-shielded arc welding, but these processes are inoperative on such metals as magnesium, or on very thin aluminum castings.

A variant of inert-gas-shielded arc welding which is becoming increasingly important in steel foundries is *metal-inert gas* welding. In this process, bare wire of the proper composition is used for welding rod; it also serves as the electrode. The wire is fed automatically from a large spool to the work, and an inert gas shield protects the wire during welding. Argon, nitrogen, or carbon monoxide is used as the shielding gas. Metal-inert gas welding permits very high production rates but is usually economical only when full utilization can be made of the speed of the process; it is finding use for repair and joining of large steel castings.

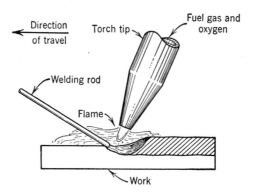

FIG. 14.20. Gas welding.

14.18 Gas welding

In gas welding the heat of chemical combustion of two gases is used to melt the surfaces to be joined (Fig. 14.20). Welding of some of the lower-melting-point metals can be accomplished by burning oxygen and hydrogen (*oxyhydrogen* flame), but for most commercial metals welding is conducted with a mixture of *oxygen* and *acetylene*.

The oxyacetylene flame (Fig. 14.21) is particularly well adapted to welding. Flame temperatures as high as 5500 to 6000°F (3100 to 3300°C) are obtained, and the chemical characteristics of the flame and its action on the weld metal can be regulated closely.

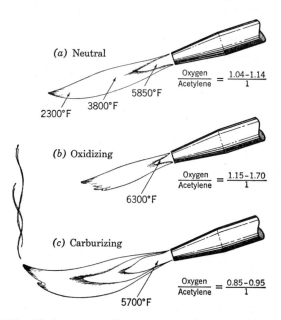

(a) Neutral

$$\frac{Oxygen}{Acetylene} = \frac{1.04-1.14}{1}$$

5850°F

3800°F

2300°F

(b) Oxidizing

$$\frac{Oxygen}{Acetylene} = \frac{1.15-1.70}{1}$$

6300°F

(c) Carburizing

$$\frac{Oxygen}{Acetylene} = \frac{0.85-0.95}{1}$$

5700°F

Fig. 14.21. Various types of oxyacetylene flames used in welding.

The complete combustion of acetylene is represented by the chemical equation

$$2C_2H_2 + 5O_2 = 4CO_2 + 2H_2O$$

$$+ \text{ Heat} \quad (542,700 \text{ Btu per lb-mole } C_2H_2)$$

For complete combustion, $2\frac{1}{2}$ volumes of oxygen are needed for each volume of acetylene. In practice a ratio of 1 volume of oxygen to 1 volume of acetylene is used. Incomplete combustion occurs at the tip of the torch, producing the brilliant light-blue cone of the flame according to the reaction

$$C_2H_2 + O_2 = 2CO + H_2 + 193,000 \text{ Btu}$$

The CO and H_2 produced now burn in air to form the outer envelope of the flame. The reactions are

$$2CO + O_2 = 2CO_2 \quad \text{and} \quad H_2 + \tfrac{1}{2}O_2 = H_2O$$

The net result is the so-called *neutral flame* (Fig. 14.21*a*), which develops only about ⅓ of the total heat available.

This neutral flame is ideal for welding steel; the molten pool of weld metal remains clear and quiet with no boiling, foaming, or sparking. The flame protects the metal from oxidation since it is neutral. The result is clean, sound, and ductile weld metal if all other variables such as welding rod and manipulation are under control.

Now, if the gas ratio is altered to give 1.15–1.5 parts of oxygen to 1 part of acetylene, an *oxidizing* flame is produced (Fig. 14.21*b*). This flame is short and has a slightly *necked-down* inner cone. The flame is very hot, about 6300°F (3500°C), and will *burn* steel; such burning is evidenced by sparking and foaming, and the resulting welds are very brittle. It is somewhat anomalous that an oxidizing flame is used for welding copper, bronzes, and brasses because the oxide film produced protects the weld from the absorption of hydrogen.

Besides neutral and oxidizing flames, the gas mixture can be so adjusted that the *ratio of oxygen to acetylene is about* 0.85 *or* 0.95 *to* 1, and the flame produced is *carburizing* or *reducing* (Fig. 14.21*c*). A reducing flame comprises three distinct zones. This flame is not quite as hot as the neutral flame, but it is hot enough to cause the molten steel of the weld to boil and become *muddy* looking. It is chiefly used *for welding alloy steels and aluminum;* in such cases the readily oxidizable elements should be protected and a reducing atmosphere is essential.

Oxyacetylene gas is ideal *where sections to be welded are not too heavy,* and where *slower cooling rates are required* for hardenable steels to minimize the hardness of the weld. This is done by playing the flame over the metal and preheating the area ahead of and around the weld; such preheating can be done by the broad flame of the gas torch where it cannot be accomplished by the arc in electric welding. Further, the lower temperature of the flame (compared to the arc) requires slower progression, and thus automatically results in slower cooling rates.

14.19   Exothermic welding

Exothermic welding for repair of castings has already been described in Chapter 9. The same principle is applied to the joining of castings by exothermic welding. The main industrial use for exothermic welding is for joining heavy sections of ferrous metals. In such cases, the exothermic reaction is

$$8Al + 3Fe_3O_4 = 9Fe + 4Al_2O_3$$

$$+ \text{Heat} \quad (179,000 \text{ Btu per lb-mole Al})$$

Finely divided iron oxide and aluminum powders are used as reactants. Both the heat and the molten iron from the reaction accomplish the actual welding. A typical arrangement for such a process is shown in Fig. 14.22.

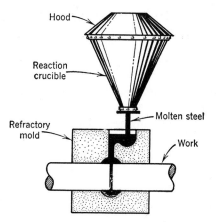

FIG. 14.22. Thermit welding.

Aside from repair work, there has been an increasing tendency to use exothermic welding in the *fabrication of heavy units;* i.e., by joining together relatively simple castings, thus avoiding the necessity of casting excessively large, intricate parts. The advantages of exothermic welding for this type of work are that (1) cost is lowered by the *elimination of many foundry operations and extensive machining,* (2) *time consumed in the operation* is less in comparison with other methods, particularly arc welding, and (3) the operation relieves stresses and no additional stress relief is required.

14.20 Metallurgical difficulties in the weld deposit

In fusion welding, a molten weld deposit is formed at the junction of the two joined surfaces. The weld bead solidifies like a rapidly chilled casting, and it is subject to the same metallurgical problems as might be expected in such a casting.   Figure 14.23 illustrates schematically the solidification of a weld cross section.

Dissolved gases are a constant problem if welding conditions are not sufficiently controlled.   Gases may be absorbed from the atmosphere or electrode coatings and precipitate during solidification to form *blowholes*.   Sometimes the rapid solidification prevents precipitation of gas, which then remains in the metal in a supersaturated condition.   When nitrogen is the supersaturated gas, a hard compound called *nitrogen-pearlite* or *braunite* tends to form.   Braunite seriously *embrittles* steel welds.   Hydrogen, when supersaturated, diffuses to grain boundaries and into crystal lattice defects, causing *hydrogen embrittlement*.

Gas absorption by weld deposits can be reduced or prevented by the selection of proper welding procedures.   By using flux-coated electrodes, a pool of slag is deposited over the molten metal continuously as the weld progresses.   If the slag chosen is dry and does not contain harmful organic binders, absorption of gases from the atmosphere can be prevented.   Overheating during welding should be avoided, as gas absorption is proportionately more

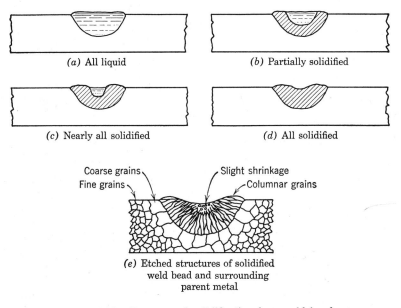

(a) All liquid

(b) Partially solidified

(c) Nearly all solidified

(d) All solidified

Coarse grains
Fine grains
Slight shrinkage
Columnar grains

(e) Etched structures of solidified
weld bead and surrounding
parent metal

Fig. 14.23.   Progress of solidification in a weld bead.

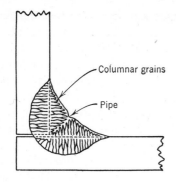

Fig. 14.24. Development of pipe in a fusion weld.

rapid at high temperatures and metallurgical properties may be impaired.

With slag-coated electrodes there is some danger of trapping slag in the weld bead. The slag immediately beneath the arc is churned vigorously into the molten metal; fortunately, if the temperature of the weld deposit is adequate, enough time elapses before solidification to permit the slag to float to the top as the arc moves on, where it coalesces into a solid layer. With poor welding technique, particularly when the weld metal lacks fluidity, some of this slag may be trapped as inclusions by the growing columnar crystals. Usually such globular inclusions are not as serious as the film of oxide that sometimes forms when the electrode is melted drop by drop with inadequate flux. If the weld is not stirred (puddled) enough to work the films of oxide to the surface, a damaging inclusion remains in the weld. Serious slag inclusions may result if a weld bead is laid upon a previous bead (multiple-pass welding) without completely chipping and brushing the slag coating from the previous bead. These relatively large inclusions seriously impair the strength of the welded joint.

Oxide films and gas absorption can be prevented by proper welding technique and the use of fluxes. Slag inclusions can be prevented by proper control of welding temperature which determines the viscosity of the melt, and by agitation of the molten pool. Cleaning of previous welds in multiple-pass welding, choice of proper fluxes, and control of welding rate are also important considerations in preventing slag entrapment.

Shrinkage cavities can occur in welding; they resemble the pipes so typical of castings. Figure 14.24 shows a large single-pass fillet weld. Solidification proceeds from both sides at the same rate, and the growing columnar crystals meet at the center. Since solidification is so rapid, molten metal cannot be supplied to compensate for shrinkage along the interface, and a plane of weakness develops. This can be prevented by depositing several small beads rather than one large one.

Sometimes welds crack at or near the interface between bead and parent metal. This can be due to exceptionally high stresses resulting from bad weld design, from transformations in metals not suitable for the welding procedure used, or *red-shortness* (hot tearing) in the steel brought about by excessive sulfur.

14.21 Metallurgical difficulties in the heat-affected zone

Adjacent to the weld bead (molten zone) of each weld lies a zone of the original base metal which was not melted but which was heated to very high temperatures by the welding operation and quickly chilled. This is the weld *heat-affected zone*. Figure 14.25 illustrates the temperature distribution across the heat-affected zone of a welded 0.80 per cent carbon steel. Over a relatively narrow region, the temperature increases from room

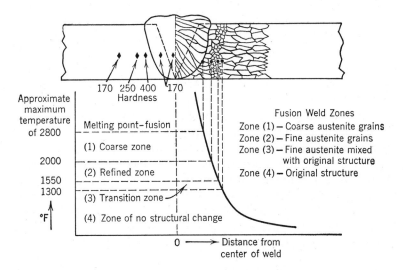

Fig. 14.25. Survey of hardness and microstructure across a welded steel plate.

temperature to above the melting point of steel at the weld joint. It is a characteristic of welds that, because of this steep thermal gradient, cooling after welding is extremely rapid; unless very special precautions are taken, the heat-affected zone behaves as if it were given a rapid *quench* after welding.

If the steel to be welded is one that is *hardenable* as is the 0.80 per cent carbon steel, the "quench" from high temperatures will produce a martensitic structure near the weld. The structure is extremely hard as shown in Fig. 14.25; it is also brittle and, when much martensite is present in a weld, cracking is likely to occur. Cracking is particularly apt to occur under such conditions if hydrogen is dissolved. Preheating is often used for hardenable steels to minimize martensite formation. The preheating reduces the cooling rate at the weld and favors pearlite formation.

In welding non-ferrous castings, the heat-affected zone is not usually as great a problem as with steel. If age-hardenable alloys are welded in the heat-treated condition, a brittle overaged zone can result, making further treatment desirable.

14.22   Stresses and distortion in welding

The thermal effects of welding can result in physical, as well as metallurgical, changes in a fabricated part. Upon heating, metals expand; they also become weaker and more ductile, and so may flow *plastically*. On subsequent cooling, they shrink and regain their former strength. As a result, welding causes *internal stresses* which may warp the casting, or which may be great enough to cause cracks to form. Figure 14.26 illustrates how welding can cause warpage, if conducted improperly.

Many techniques are available to the welder to minimize stresses. These include control of the size and shape of weld bead, and the number of passes made. Welds are often *peened* between passes to minimize stresses. *Preheating* has already been mentioned as a method for reducing the tendency toward hardening in a weld; it also reduces internal stresses. *Interpass temperature* (the temperature of the work between passes) must also be controlled. Finally, *postheating* is often used to temper any transformation products and to stress relieve the weld.

14.23   Weldability

"Weldability" is a difficult term to define accurately, but one that is widely used to refer to the ease with which a given alloy

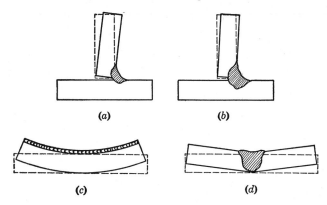

Fig. 14.26. Permanent distortion developed during cooling after welding. (Dotted lines show original position.) (*a*) Distortion caused by metal shrinkage when members are unrestrained (slight penetration in fillet weld). (*b*) Less distortion caused by shrinkage when deeper penetration weld is used. (*c*) Distortion caused by welding one surface of plate. (*d*) Distortion caused by welding V joint from one side.

can be welded.   It is best thought of in terms of the precautions necessary to obtain a sound, crack-free weld of requisite strength and ductility.   In general, plain-carbon and low-alloy steels containing less than about 0.3 per cent carbon are easily weldable, because little or no martensite is formed in the heat-affected zone. Steels of greater carbon content are more hardenable and, therefore, require care in preheating and postheating to assure useful welds.   Cast iron is not easily weldable since it is inherently brittle; also, white iron tends to form because of the rapid cooling entailed unless the part is carefully preheated and postheated.   Special welding rods and careful techniques must be used to obtain satisfactory welds on cast irons.   Some non-ferrous alloys are weldable only with difficulty because of hot tearing, reactivity, or other reasons.

The welding process, like the casting process itself, is more adaptable to some alloys than to others.   With careful techniques, however, a wide range of alloys can be joined successfully by welding, and with each new development in the still young science of welding, its usefulness to the foundryman increases.   The various welding handbooks [7, 8] should be consulted for specific welding techniques and precautions.

14.24　Summary

For many castings, *heat treatment* has become a vital step between the casting production and its use as an engineering part. Cast parts are heat-treated for one or more of the following reasons: to stress-relieve, soften, harden, homogenize the structure, refine the grain, or alter mechanical properties in various ways.

Stress-relief treatments do not alter the casting structure but are intended only to relieve internal stresses to prevent failure or warping in subsequent machining operations or in service. High-strength aluminum and magnesium alloys are heat-treated by *solution treating, quenching,* and *aging.* The solution treatment homogenizes the structure, quenching retains the homogeneous structure at room temperature, and aging strengthens the material by causing fine precipitates to form in the structure. Most copper-base alloys are not heat-treated (except for stress relief). Two exceptions are beryllium-bronze and aluminum-bronze, both of which are given high-temperature homogenizing treatments, quenched, and aged (tempered).

Cast steels are usually *annealed, normalized,* or *quenched and tempered.* Annealing involves heating to the austenite region (for hypoeutectic steels) and furnace cooling. *Annealing* homogenizes the structure, refines the grain, and produces a relatively soft and machinable structure. A stronger part is produced by *normalizing* (air cooling from the austenite region). Still higher strengths with good ductility are obtained by *quenching* and *tempering.*

Gray, malleable, and ductile cast irons can be heat-treated in the same fashion as steel to improve mechanical properties and machinability, or for other reasons. Heat treatments do not improve the strengths of gray iron to any appreciable degree because of the presence of embrittling graphite flakes; they do, however, increase mechanical properties of malleable and ductile iron. Heat treatments are used for all types of cast iron to improve machinability and wear resistance.

Just as heat treatment has become a necessary adjunct to successful foundry operation, so also has *welding.* Welding is used to build up relatively simple castings into complex *cast-weld* structures; it is also used to join castings and forgings into *composite* structures, and to *repair* castings. Many different welding processes are in use today in foundries. These have been discussed in detail in this chapter; they include *gas* welding, *arc* welding, and *thermit* welding processes. With careful procedures and controls, welding can produce a metal joint fully as sound and strong as the base casting. Proper use of welding can result in tremendous savings in the foundry, and can permit production of parts which are more complex than can be readily cast in one piece.

14.25 Problems

1. With a simple diagram, illustrate the type of binary alloy system responsive to age hardening. List a number of technical alloy examples. How will coring affect the age-hardening process?

2. Why should age-hardenable aluminum used in structures present welding difficulties?

3. Gray iron castings were formerly stress-relieved by "aging" them in the yard for several years. Was this effective? Is there a better way to accomplish this?

4. How can you refine the grain size of steel by annealing? What grain is refined?

5. What information is made available by the *TTT* curve that was lacking in the iron-carbon equilibrium diagram?

6. Of what operations does (*a*) the hardening of steel usually consist? (*b*) the annealing? (*c*) the normalizing? What is the purpose of each operation?

7. Define hardenability. How is it related to the critical cooling rate? How is the hardenability of carbon steel affected by carbon content and grain size? In general, how do alloying elements affect hardenability?

8. If a steel possesses high hardenability, does this necessarily mean that the steel is extremely hard in the quenched condition? Explain your answer.

9. What heat treatments can be used for ductile iron to increase the tensile and yield strength? How would these alter the microstructure?

10. Would the heat treatments of Problem 9 alter the microstructure of gray cast iron in similar manner? Would the mechanical properties change in similar manner?

11. Obtain samples of (*a*) cast 0.30 plain carbon steel, and (*b*) cast 0.30 carbon steel containing additions of Cr–Ni–Mo for improved hardenability.

Cut specimens (preferable 1-in.-diameter round cylinders) from each of the steels. Choose an acceptable austenitizing temperature, and perform the following treatments on specimens of each steel:

(*a*) air-cool, (*b*) oil-quench, (*c*) water-quench. Temper all specimens at 400°F for 1 hour.

Section the cylinders along their diameter, and measure depth of hardening. Present your data and write a short discussion of the relationship between hardenability, hardness, and mechanical properties in 0.30 carbon-steel castings of (*a*) 1 in. section thickness, and (*b*) ⅛ in. section thickness.

12. What welding process would you be most likely to use to join castings of (*a*) magnesium alloys, (*b*) aluminum alloys, (*c*) gray iron, (*d*) steel?

13. State three reasons for using a coated metal electrode instead of a bare steel wire in arc welding.

14. What kind of a substance is an "arc stabilizer"?

15. Write the chemical reactions between oxygen and acetylene for complete combustion as used in gas welding.

16. How does a neutral flame differ from a reducing flame when oxyacetylene gas mixtures are used?

17. How would you weld copper with a gas torch: i.e., what type of flame would you choose, and why?

18. Which gives a hotter flame, oxygen plus acetylene or oxygen plus hydrogen?

19. Why is it so easy to cut steel with a gas mixture of oxygen and acetylene? Write the chemical reaction.

20. Why should there be boiling and foaming when the gas flame is oxidizing?

21. List two advantages of gas welding, and discuss them briefly.

22. Write the chemical reactions involved in thermit welding. When may it be used to fullest advantage?

23. What is the difference between a "blowhole" and a "pipe" in a solidified weld? Indicate how they originate.

24. What factors favor hot cracks while the weld metal is in the mushy stage?

25. What connection is there between the "hardenability" of a steel and its "maximum safe cooling rate" in welding? How would you slow up the cooling rate of a steel in welding?

26. The electric arc is much hotter than the oxyacetylene flame. Why might a steel not weldable with the arc be weldable when torch-welded?

27. "Flow-weld" two aluminum castings together by ramming them close together in a sand mold and cutting a passage way through the sand to allow 10 to 20 lb of aluminum to flow past the joint.

Section the welded castings, and macroetch them. Is the structure you obtain what you would have expected? Why is it that properly made welds of this type can be stronger than the base metal, although chemical analysis and heat treatment of the weld and the base metal may be the same?

28. Make a list of factors that might lead you to adopt cast-weld structure in preference to making a large single piece casting.

29. Obtain a hardenable (as-cast) steel casting, and make a weld bead by melting a small pool on the casting surface.   After welding, temper at 800°F.

Sketch the microstructure you expect to be present in and near the weld bead.   Examine a series of microstructures in the weld area to check your predictions.

30. Repeat Problem 29 for aluminum–4.5 per cent Cu (age-hardenable) alloy.

14.26   Reference reading

1. *A.S.T.M. Standards,* American Society for Testing Materials, Philadelphia.

2. *Metals Handbook,* A.S.M., Cleveland.

3. *Atlas of Isothermal Transformation Diagrams,* U. S. Steel Co., Pittsburgh, 1951.

4. *Atlas of Isothermal Transformation Diagrams of B. S. En Steels,* second edition, Iron and Steel Institute, London, 1956.

5. Briggs, C. W., *The Metallurgy of Steel Castings,* McGraw-Hill, 1946.

6. *Cast Metals Handbook,* fourth edition, A.F.S., Des Plaines, 1957.

7. *Welding Handbook,* A.W.S., New York, 1950.

8. *Procedure Handbook of Arc Welding Design and Practice,* Lincoln Electric Co., Cleveland, 1950.

9. Henry, O. H., and Clausen, G. E., *Welding Metallurgy, Iron and Steel,* A.W.S., New York, 1949.

10. Rominski, E. A., and Taylor, H. F., "Stress Relief and the Steel Casting," *Transactions A.F.A.,* **51**, 709–736, 1943.

Index